Contents

Revision Tips

What ...

◎ Use this Revision Guide and your exam specification to see what topics you need to cover. Ask your teacher which specification you are following and download one from the web.

◎ Go through the topics, any past papers and tests and make a list of the areas that you find difficult. Concentrate your revision on these.

◎ Check out which formulae will be given on the exam paper. Your teacher can tell you this. Practise using these and make sure you learn any formulae that aren't given.

When and where ...

◎ Start your revision early. Aim to start at least a month or two before your exam, preferably earlier.

◎ Revise little and often rather than in one long session. Study for no more than 45 minutes and then take a 10 minute break.

◎ If your school or college has a revision club then check it out.

◎ Find a quiet, well lit place to study and have all your books handy. Keep some scrap paper nearby to write notes and practise questions.

◎ Avoid distractions such as the television, although if you find working to music helps you to concentrate then use it.

◎ Don't revise late at night or very early in the morning – it's important to be alert.

◎ Eat well, take some exercise and get enough sleep. When you're away from your revision enjoy yourself and don't think about work.

How ...

◎ Create an effective revision timetable.

- Find out the date of your exam.
- Divide your subjects up into topics.
- Mix and match harder and easier topics to break it up a bit.
- Schedule 10 minutes of top up time at the start of each session to look back over the stuff you covered last time.
- Tick off topics as you've done them.
- Timetable in some time off and rewards.
- Allow time in the last week to do exam style questions and go over everything one last time.
- Stick to your schedule but if you feel comfortable with some topics and are struggling with others, shift it around to allow extra time on the harder topics.

◎ Use cards to summarise the key points listed in this book. Condense them to one side of paper and take it everywhere with you, reading it at every opportunity.

◎ Use mnemonics. For example, Never Eat Soggy Waffles helps you remember the points of the compass for bearing questions. Illustrate your key points with diagrams and use highlighter pens.

◎ Record yourself reading out key points or formulae and play them back later.

◎ Write formulae on sticky notes and put them where you'll see them regularly – the fridge, the bathroom mirror, ...

◎ Get a revision buddy. Take it in turns to explain topics to each other. Test each other. Discuss what works and doesn't work.

AQA Mathematics
for GCSE

Exclusively endorsed and approved by AQA

Revision Guide

Series Editor
Paul Metcalf

June Haighton
Andrew Manning
Kathryn Scott

FOUNDATION
Linear   Modular

Nelson Thornes

Published in 2007 by:
Nelson Thornes Ltd
Delta Place
27 Bath Road
CHELTENHAM
GL53 7TH
United Kingdom

08 09 10 11 / 10 9 8 7 6 5 4 3 2

A catalogue record for this book is available from the British Library

ISBN 978 0 7487 8194 2

Cover photograph by Oxford Scientific Films
Illustrations by Heelstone Publishing Services and Roger Penwill
Page make-up by Heelstone Publishing Services

Printed and bound in Croatia by Zrinski

- Create mind maps or spider diagrams for different topics using plenty of colour and stick them on your wall.

- Don't try to memorise maths; try to understand the processes.

- Don't be tempted to spend too long on topics that you find easy; concentrate on things that you find harder or get wrong.

- Use the end-of-chapter questions to make sure you've understood everything.

- Use the exam style questions to practise, practise, practise.

Exam Tips

- Take the correct equipment. Two pens (only write in blue or black), two sharp pencils, a pencil sharpener, a ruler, a rubber, your calculator with spare batteries, protractor, compasses and a watch to time your answers.

E I 9 h t

- Things you need to know about your calculator before you sit your exam:
  - check that your calculator follows BODMAS rules
  - how to use brackets
  - how to use the memory
  - how to use the Ans button to insert the previous answer
  - how to find a square root ($\sqrt{}$)
  - how to find a power of a number (32, 5^4, ...)
  - how to do fractions
  - how to enter and read figures in standard form

Not all calculators have this function

Try typing 20 – 9 × 2 = , if your calculator gives the answer 2 it's following the rules. If it gives the answer 22 it's not and it's worth getting a newer calculator.

- Read through the paper for a few minutes to allow your nerves to settle.

- Write your mnemonics down against relevant questions as a reminder.

- Lower graded, easier questions are at the beginning of the paper so start with those to ease your way in.

- Pace yourself so that you don't run out of time and try to allow 10 minutes at the end for reading through your answers.

- Try to work neatly and set out your answer clearly in the given space.

- Look at the marks available for a question. A question worth 1 mark doesn't need lots of explanation or working, but a 4 mark question is likely to need more than one step to get the answer.

- Show all your working for questions worth more than 1 mark. If you simply write down the answer you are gambling that it's completely correct as the examiner won't be able to award you part marks when your method isn't there.

- Use appropriate calculator methods on the calculator paper. For example, build up and long multiplication are methods that are best suited to the non-calculator paper.

🌀 Read the question carefully and interpret the 'exam speak':

Exam speak	What it means
Write down	Working isn't needed.
You must show your working/Explain your answer/Give a reason for your answer	Show your method; you will not get any marks otherwise. You can use words, numbers or algebra.
Estimate	Round the numbers, then do the calculation. Don't find the exact answer.
Work out/Calculate/Find	Do a calculation; don't measure.
Measure	Use a ruler or protractor; don't calculate.
Not drawn accurately	The diagram is not accurate; don't measure it.
Use the graph to estimate/solve	Your answer must come from the graph.
Give your answer to a suitable degree of accuracy	Use the same or less accuracy than the numbers given in the question.
State the units of your answer	There will be a special mark for this.

🌀 Check your answer is sensible. For example, if it is a probability is it between 0 and 1?

🌀 If the question asks you to give your answer to a degree of accuracy then make sure you do.

🌀 Don't round until the end of the question (unless it's an estimating question) or you may lose accuracy.

🌀 Study the Exam style answers to see what the examiner is looking for.

Good Luck
and
Don't Panic!

Collecting data

Secondary data might include data in books, newspapers, magazines, etc or else in a database.

Key points

Classifying data

- **Primary data** is data you collect yourself.

- **Secondary data** is data that someone else has collected.

- **Qualitative data** is data that cannot be measured using numbers (for example, hair colour, favourite film, type of car, etc).

- **Quantitative data** can be measured using numbers (for example, height, weight, temperature, age, collar size, etc). Quantitative data can be split into two categories.

  - **Discrete data:** data that can only take exact numbers, for example, shoe sizes, as you can have $3\frac{1}{2}$ and 4 but nothing in between.

  - **Continuous data:** data that can take any values, for example, length, as you can have 10 cm and 11 cm and any number of measurements in between.

Sampling methods

When undertaking surveys, you will need to decide a sampling method. There are five main types of sampling methods that you should be familiar with:

Surveys might include direct observation or personal surveys (for example, face-to-face interviews or telephone, postal and internet surveys). You will need to be able to design surveys and questionnaires for the exam.

- **Convenience** or **opportunity sampling** means that you just take the first people who come along or those who are convenient to sample (such as friends and family).

- **Random sampling** requires each member of the population to be assigned a number and the sample is chosen using random numbers.

- **Systematic sampling** involves taking every nth member of the population where the value of n is chosen by dividing the population size by the sample size.

- **Quota sampling** involves choosing a sample with certain characteristics, for example, selecting 20 adult men, 20 adult women, 10 teenage girls and 10 teenage boys.

- **Cluster sampling** splits the population into smaller groups, or clusters, and takes a random selection of these clusters and surveys all the members within them.

Worked Example Classifying data

Question

Say whether each of the following are qualitative or quantitative.

Where the answer is quantitative state whether the data is discrete or continuous.

a Temperature in a room

b Number of cars at a junction

c Colours of cars at a junction

d Cost of mobile phones

It is easy to confuse qualitative and quantitative – remember that **qualitative** goes with **quality** and **quantitative** goes with **quantity**.

Solution

a Temperature in a room is quantitative and continuous

b Number of cars at a junction is quantitative and discrete

c Colours of cars at a junction is qualitative

d Cost of mobile phones is quantitative and discrete

It is also easy to confuse discrete and continuous – remember, for example, that shoe size is discrete but shoe length is continuous.

Worked Example Questionnaire design

Question

Frank is writing a questionnaire about computer games.

This is one question from Frank's questionnaire:

> How much money do you spend on computer games?
>
> Tick one of the boxes.
>
> £0–10 ☐   £10–20 ☐   £20–30 ☐

Write down two criticisms of Frank's question.

AQA ↗

EXAMINER SAYS…

When designing questionnaires keep questions short and simple, include time scales, avoid sensitive issues and do not ask personal questions.

Remember that the best questionnaires are:
- **appropriate** to the survey being carried out and do not ask unnecessary questions
- **unbiased** so they do not lead you to give a particular answer
- **unambiguous** so they are clear and straightforward.

Solution

The question does not give a time scale so it is not clear whether it is per week, per month etc.

The intervals overlap so it is not clear where £10 or £20 would go.

Worked Example Representing data (1)

Question

The following information shows the numbers of brothers and sisters in a form group.

0	5	0	1	1	5
1	2	1	2	2	3
1	1	2	2	1	2
1	0	6	0	1	0

Complete a tally chart and a frequency distribution for this data.

In this form, the information is called <u>raw data</u> because it has still to be organised.

Solution

Number of brothers and sisters	Tally	Frequency
0	IIII	5
1	IIII IIII	9
2	IIII I	6
3	I	1
4		0
5	II	2
6	I	1

24 ←

Remember that tallies are grouped in fives so that

IIII = 4   IIII = 5

IIII I = 6 etc

This makes the tallies easier to read.

 **GET IT RIGHT!**

Always add up the frequencies to check you have the correct total.

 **BUMP UP**

C

THE GRADE   To get a grade **C** you will need to work with grouped tally charts, but make sure you put the entry in the correct group so 50 goes in the $45 < h \le 50$ group not the $50 < h \le 55$ group.

Worked Example Representing data (2)

Question   Another form group records their information as a two-way table.

Brothers and sisters

	0	1	2	3	4	5	6
Girls	3	5	4	2	1	0	0
Boys	2	3	3	0	1	0	1

A two-way table is useful to identify different groups.

The total for the rows and the total for the columns should be the same – you may wish to add these to the table.

a How many boys are there altogether in the class?

b How many girls have four brothers and sisters altogether?

c What fraction of the class has no brothers and sisters?

d What is the maximum number of children in any family in the form group?

Solution   a There are 10 boys altogether

b 1 girl has four brothers and sisters altogether

c Fraction of the class is $\frac{5}{25} = \frac{1}{5}$

d 7

GET IT RIGHT!

Think carefully about the wording of questions such as this. The table says that 1 boy has 6 brothers and sisters altogether so there are 7 children in his family.

Collecting data

END OF CHAPTER QUESTIONS

Time Yourself!

Can you complete these questions in **30** minutes?

1 For each of the following say whether the data is quantitative or qualitative.

   a The number of people at a rugby match.

   b How many tins of beans a shop sells.

   c The flavour of the beans.

   d The time it takes to travel from London to Manchester.

2 For each of the following say whether the data is discrete or continuous.

   a The number of votes for a party at a local election.

   b The number of beans in a tin.

   c The weight of a tin of beans.

   d The time taken to complete this chapter.

3 The two-way table shows the colour and make of cars in a car park.

	Red	Blue	Black	White
Ford	3	5	1	2
Vauxhall	2	4	3	0
Toyota	1	2	0	2
Other	2	2	3	3

Use the table to answer the following questions.

   a How many red cars were there in the car park?

   b How many Vauxhall cars were there in the car park?

   c What percentage of the cars were black?

4 A college wishes to undertake a survey on its sports facilities.

   Explain how you would take:

   a a random sample of 50 students

   b a systematic sample of 50 students.

2

Statistical measures

Key words

mean

mode

modal class

median

range

grouped data

frequency table

frequency distribution

Key points

If there are two middle values, add the two values together and then divide by two.

◎ The **mode** is the value that occurs most often.

◎ The **median** is the middle value when the values are arranged in **order of size**.

◎ The **mean** is found by calculating $\dfrac{\text{the total of all the values}}{\text{the number of values}}$

◎ For a **grouped** distribution, the **mean** is found by calculating $\dfrac{\text{the total of all the (frequencies} \times \text{midpoint values)}}{\text{the total of frequencies}}$

or the mean $= \dfrac{\Sigma fx}{\Sigma f}$ where Σ means 'the sum of'

◎ The **range** is the difference between the largest and smallest numbers.

Worked Example — Mean, median, mode and range

Question

A group of friends had these marks in a test.

6   9   25   30   30   35   25   30   25   30   30

Work out the mean, median, mode and range of their results.

Solution

$\text{Mean} = \dfrac{\text{the total of all the values}}{\text{the number of values}}$

$= \dfrac{6 + 9 + 25 + 30 + 30 + 35 + 25 + 30 + 25 + 30 + 30}{11} = \dfrac{275}{11} = 25$

Median is the middle value when arranged in order of size

Arranging in order   6   9   25   25   25   30   30   30   30   30   35

Middle number, so median = 30

30 appears 5 times

The mode is the value that occurs most often, so mode = 30

The range is the difference between the largest and smallest numbers = 35 − 6 = 29

GET IT RIGHT!

When finding the median, you must remember to put the numbers in order of size.

GET IT RIGHT!

The range should always be given as a single number not a range, so 6 – 35 or 35 – 6 is not correct.

Worked Example Frequency distributions

Question

The number of students in 25 classes is shown in the table below.

Number of students in a class (x)	Frequency (f)
25	1
26	2
27	4
28	4
29	6
30	4
31	2
32	2

Work out the mean, mode, median and range of the data.

Solution

Completing the table:

AQA EXAMINER SAYS...

The examination paper will leave sufficient space for you to add further columns. Don't waste time copying out the table again.

Number of students in a class (x)	Frequency (f)	Frequency × students (fx)
25	1	25 × 1 = 25
26	2	26 × 2 = 52
27	4	27 × 4 = 108
28	4	28 × 4 = 112
29	6	29 × 6 = 174
30	4	30 × 4 = 120
31	2	31 × 2 = 62
32	2	32 × 2 = 64
Total	$\sum$**f = 25**	$\sum$**fx = 717**

Start by adding an extra column to work out 'fx'

Complete the totals to give $\sum$f and $\sum$fx

$\sum f = 1 + 2 + 4 + 4 + 6 + 4 + 2 + 2$

$\sum fx = 25 + 52 + 108 + 112 + 174 + 120 + 62 + 64$

Mean $= \dfrac{\sum fx}{\sum f} = \dfrac{717}{25} = 28.68$

GET IT RIGHT!

The mode is 29 as this is the value that occurs **most** often.
In this example, number 6 is the largest frequency.

Mode $= 29$

If there are n values then the middle value is the $\dfrac{(n + 1)th}{2}$ value.

The median is the **middle** value when arranged in order of size.

Since there are 25 values then the median is the $\dfrac{(25 + 1)th}{2} = 13$th value.

Number of students in a class (x)	Frequency (f)	Cumulative frequency	
25	1	1	=1
26	2	1+2	=3
27	4	1+2+4	=7
28	4	1+2+4+4	=11
29	6	1+2+4+4+6	=17
30	4		
31	2		
32	2		

GET IT RIGHT!

The range should always be given as a single number. Make sure you subtract the values (32 – 25) and not the frequencies (6 – 1).

If you add up the frequencies you get the cumulative frequency (Σf)

For example, Σf = 3 tells you that, altogether, there are 3 classes with up to 26 students in them.

Σf = 11 means there are 11 classes with up to 28 students in them.

Median = 29

Range 32 – 25 = 7

The 13th value is in this group. (29 students per class) so 29 is the median.

THE GRADE To get a grade **C** you must be confident working with grouped data and finding an estimate of the mean of grouped data. Practise these types of questions in your revision programme.

Worked Example — Group frequency distributions

(Question) The heights (to the nearest cm) of 80 students are given in the table below.

EXAMINER SAYS…

On the examination paper, groups can be written in different ways such as 100 up to 110, $100 \leqslant x < 110$, 100 + ...

Height in cm	Number of students
150–154	3
155–159	4
160–164	9
165–169	16
170–174	18
175–179	17
180–184	7
185–189	6

a What is the modal class of the distribution?

b What is the estimated mean height of the students?

c Which class interval contains the median?

Solution

Completing the table:

Start by adding extra columns for the midpoint at the mid-interval value and to work out 'fx'

Complete the totals to give $\sum f$ and $\sum fx$

Height in cms	Number of students	Midpoint (x)	Midpoint × frequency (fx)
150–154	3	152	152 × 3 = 456
155–159	4	157	157 × 4 = 628
160–164	9	162	162 × 9 = 1458
165–169	16	167	167 × 16 = 2672
170–174	18	172	172 × 18 = 3096
175–179	17	177	177 × 17 = 3009
180–184	7	182	182 × 7 = 1274
185–189	6	187	187 × 6 = 1122
Total	$\sum f = 80$		$\sum fx = 13\ 715$

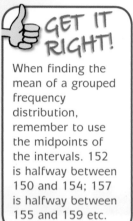

GET IT RIGHT!

When finding the mean of a grouped frequency distribution, remember to use the midpoints of the intervals. 152 is halfway between 150 and 154; 157 is halfway between 155 and 159 etc.

$\sum f = 3 + 4 + 9 + 16 + 18 + 17 + 7 + 6 = 80$
$\sum fx = 456 + 628 + 1458 + 2672 + 3096 + 3009$
$\qquad\quad + 1274 + 1122 = 13\ 715$

You cannot find the mode of a grouped frequency distribution, but the modal class is the interval with the highest frequency.

a The modal class is 170–174 cm

b Mean $= \dfrac{\sum fx}{\sum f} = \dfrac{13\ 715}{80} = \ = 171.4375 = 171$ cm (3 s.f.)

c The median is the middle value when arranged in order of size.

Since there are 80 values, then the median

is the $\dfrac{(80 + 1)\text{th}}{2} = 40.5$th value

The 40.5th value is exactly halfway between the 40th value and the 41st value.

Height in cm	Number of students (f)	Midpoint (x)	Cumulative frequency (fx)	
150–154	3		3	= 3
155–159	4		3 + 4	= 7
160–164	9		3 + 4 + 9	= 16
165–169	16		3 + 4 + 9 + 16	= 32
170–174	18		3 + 4 + 9 + 16 + 18	= 50
175–179	17			
180–184	7			
185–189	6			

The 40.5th value is in this group, which contains the 40th and 41st values.

The 170–174 class interval contains the median height.

Statistical measures

END OF CHAPTER QUESTIONS

Time Yourself!

Can you complete these questions in **25** minutes?

1 For each set of data find the mean, median, mode and range.

 a 6 10 3 5 5 5 7 3 1

 b 26 30 23 25 25 27 23 21

 c 0 1 1 2 2 2 3 3 3

2 A teacher records the number of circuits completed by 11 students as follows:

 19 10 17 18 19 18 17 22 7 13 18

 Calculate the mean, median, mode and range.

3 Becky asks 15 people how many people are in their family.

 Her results are shown in the table

Number in family	2	3	4	5	6	7	8
Frequency	1	3	6	3	1	0	1

 From this information calculate the median, mode, mean and range.

4 The following table shows the pocket money for students in a class.

Pocket money (£)	0–10	10–20	20–30	30–40
Frequency	7	9	6	3

 a What is the modal class of the distribution?
 b Calculate an estimate of the mean pocket money.

3 Representing data

Key words

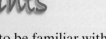

pictogram

pie chart

bar chart

stem-and-leaf diagram

line graph

time series

Key points

You will need to be familiar with the following types of representations:

Pictogram

In a pictogram, the frequency is shown by a number of identical pictures.

Colour	Frequency
Silver	📱📱📱📱📱📱📱
Black	📱📱📱📱📱
Blue	📱
Red	📱📱
Other	📱📱

Key: 📱 = 2 mobiles

Bar chart

In a bar chart, the frequency is shown by the height (or length) of the bars.

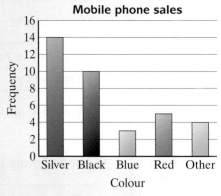

Mobile phone sales

Pie chart

In a pie chart, the frequency is shown by the angles (or areas) of the sectors of a circle.

Mobile phone sales

(pie chart showing sectors: Other, Red, Blue, Silver, Black)

Stem-and-leaf diagram

In a stem-and-leaf diagram, each value is represented as a stem (tens in this example) and leaf (units in this example).

Number of minutes to complete homework

Stem (tens)	Leaf (units)
1	8 9 9
2	3 5 7 8 9
3	3 7 8

In this case, the number 7 stands for 37 (3 tens and 7 units) that is Key 3|7 represents 37

Time series

A time series is a line graph showing values over time. A time series can be used to predict trends.

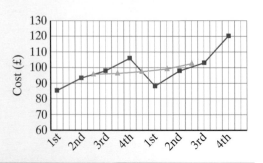

Key:

— ■ — ■ — ■ — time series

— ▲ — ▲ — ▲ — trend line

Worked Example Pie charts

Question

In a survey, holidaymakers are asked to write down their favourite method of transport. The information is shown in the following table.

Transport	Car	Coach	Train	Boat	Plane
Frequency	20	22	19	34	25

Show this information as a pie chart.

A pie chart is a representation for showing data.

Solution

In a pie chart, the frequency is represented by the angles (or areas) of the sectors of a circle.

The pie chart needs to be drawn to represent 120 people.

There are 360° in a full circle so each person will be shown by 360° ÷ 120 = 3°

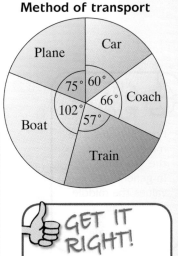

Method of transport

Transport	Frequency	
Car	20	20 × 3° = 60°
Coach	22	22 × 3° = 66°
Train	19	19 × 3° = 57°
Boat	34	35 × 3° = 102°
Plane	25	25 × 3° = 75°
Total	120	360°

AQA

EXAMINER SAYS...

In most cases, the question will provide a circle with the centre marked so that you can draw the pie chart on this. Most errors on these questions involve the use of the protractor so practise drawing angles with your protractor as these questions are usually worth a few marks.

GET IT RIGHT!

Remember to check that the sum of the angles do add up to 360°

GET IT RIGHT!

Remember to label the sectors or provide a key. Angles marked at the centre are also helpful in giving additional evidence of proportional size.

Worked Example Stem-and-leaf diagram

Question

The following information shows the number of minutes (to the nearest minute) taken to complete a puzzle.

15 13 23 8 15 23 18 9 28 6 15 22 7

a Show this information as a stem-and-leaf diagram.

b Use your diagram to find:

i the shortest time taken to complete the puzzle

ii the modal time taken to complete the puzzle

iii the median time taken to complete the puzzle.

Solution

a **Number of minutes to complete a puzzle**

Stem (tens)	Leaf (units)
0	8 9 6 7
1	5 3 5 8 5
2	3 3 8 2

In this case, the number 8 stands for 18 (1 ten and 8 units)

In this case, the number 8 stands for 28 (2 tens and 8 units)

Again it is important to provide a key: so 1 | 3 represents 13 minutes

b **i** The shortest time taken is 07 minutes or 7 minutes.

ii The modal time taken is 15 minutes (this occurs the most times).

iii To find the median it is important to have the data in numerical order.

This can be done by creating an **ordered stem-and-leaf diagram**.

Number of minutes to complete a puzzle

Stem (tens)	Leaf (units)
0	6 7 8 9
1	3 5 5 5 8
2	2 3 3 8

Key: 1 | 3 represents 13 minutes

The median time taken to complete the puzzle is 15 minutes.

Here the leaves (units) are arranged numerically.

GET IT RIGHT!

For 13 values, the middle value is the 7th value. In general, for n values the middle value is the $\frac{(n+1)\text{th}}{2}$ value.

THE GRADE

To get a grade **C** you should practise finding the median from a stem-and-leaf diagram.

AQA

EXAMINER SAYS...

Always include a key for your stem-and-leaf diagram. Similarly, if a stem-and-leaf diagram is provided in a question remember to check its key. Other formats of the key include:

85 | 2 for 852

5 | 99 for £5.99

4 | 11 for 4 feet 11 inches

3 | 55 for 3 hours 55 min

Representing data

END OF CHAPTER QUESTIONS

Time Yourself!

Can you complete these questions in **30** minutes?

1 The following table shows the sales of ice creams.

Ice cream	Vanilla	Strawberry	Raspberry	Mango	Other
Frequency	16	10	6	3	1

Show this information as

a a pictogram   **b** a bar chart   **c** a pie chart

2 The information below shows the heights, in metres, of shrubs in a garden centre.

1.55,   2.31,   1.11,   2.79,   3.23,   2.44,   1.82,   2.51,   3.04,   2.66,   1.93

a Copy and complete the following stem-and-leaf diagram to show this information.

The heights of shrubs in a garden centre

Stem (whole number)	Leaf
1	
2	
3	

Key: 2|51 represents 2.51 m

b Calculate the median height of shrubs in the garden centre.

3 Students at a college are asked to choose their favourite type of book.

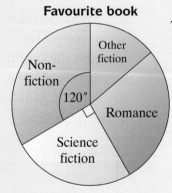

Favourite book

Their choices are shown in the pie chart below.

72 students chose 'Science fiction' books

a How many students were included in the survey?

b How many students chose 'Non-fiction' books?

Twice as many students chose 'Romance' as chose 'Other fiction'.

c How many students chose 'Romance' books?

13

4 Scatter graphs

Key words

scatter graph

positive correlation

negative correlation

strong correlation

weak correlation

line of best fit

outlier

Key points

Type of correlation

⊚ **Positive correlation**

Positive correlation is where an **increase** in one set of data happens at the same time as an **increase** in the other set of data.

For example, temperature against sales of sunglasses. As the temperature increases, the sales of sunglasses increase.

⊚ **Negative correlation**

Negative correlation is where an **increase** in one set of data results in a **decrease** in the other set of data.

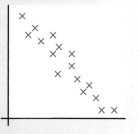

For example, temperature against sales of gloves. As the temperature increases, the sales of gloves decrease.

⊚ **Zero or no correlation**

Zero or no correlation is where there is **no obvious relationship** between the two sets of data.

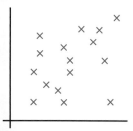

For example, temperature against sales of watches. There is no obvious relationship between temperature and watch sales.

⊚ A **line of best fit** is drawn to represent the relationship between two sets of data.

⊚ **Strong correlation**

Strong correlation is where the points lie close to a straight line. A straight line represents perfect correlation.

⊚ **Weak correlation**

Weak correlation is where the points lie roughly along a straight line but are not very close to it.

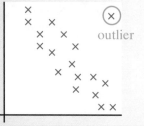
outlier

⊚ **No correlation**

No correlation is where the points are distributed across the whole diagram with no obvious pattern.

⊚ An **outlier** or **rogue value** is one that does not fit the general trend.

Worked Example   Interpreting scatter graphs

(Question) The table shows the number of days absent and the performance on an internal exam for 10 students.

Student	A	B	C	D	E	F	G	H	I	J
Days absent	1	2	2	5	15	12	24	0	8	3
Examination mark	48	47	13	42	21	32	8	49	31	35

Draw a scatter graph to show this information.

Describe the relationship shown by your scatter graph.

(Solution)

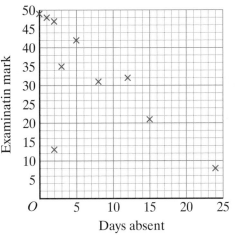

Attendance against examination

Plot your points carefully and check the scales to ensure they are all correct.

A scatter graph is a graph used to show the relationship between two sets of data. You can see from the graph that as the number of days absent increases, the examination mark decreases.

The scatter graph shows weak negative correlation between number of days absent and examination mark.

AQA

EXAMINER SAYS...

It is important that when you describe the correlation you remember to comment on the **type of correlation** *and* the **strength of correlation**. You will not gain full marks unless you make reference to both of these in your answer.

Worked Example   Lines of best fit

(Question) The table shows the average temperature and the sales of swimwear for six months in a year.

Month	Jan	Mar	May	Jul	Sep	Nov
Average temperature	6°C	9°C	12°C	18°C	15°C	10°C
Sale of swimwear	2	12	25	47	39	15

Draw a scatter graph to show this information.

Draw a line of best fit on your scatter graph.

Describe the relationship shown by your scatter graph.

Solution

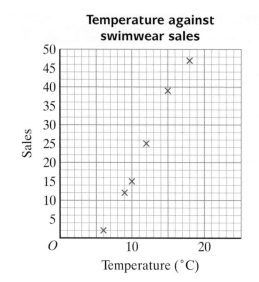

Adding a line of best fit:

A line of best fit is drawn to represent the relationship between two sets of data. A line of best fit should only be drawn where the correlation is strong.

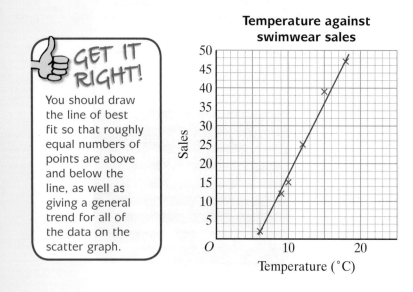

GET IT RIGHT!

You should draw the line of best fit so that roughly equal numbers of points are above and below the line, as well as giving a general trend for all of the data on the scatter graph.

For additional accuracy, the line of best fit should pass through the point $(\bar{x}, \bar{y})$ where $\bar{x}$ is the mean of all the x-values and $\bar{y}$ is the mean of all the y-values.

The scatter graph shows strong positive correlation between temperature and sales of swimwear.

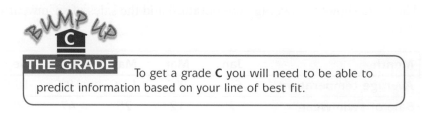

BUMP UP

C

THE GRADE   To get a grade **C** you will need to be able to predict information based on your line of best fit.

Scatter graphs

END OF CHAPTER QUESTIONS

Time Yourself!

Can you complete these questions in **45** minutes?

1 Describe the relationship shown in each of the following scatter graphs.

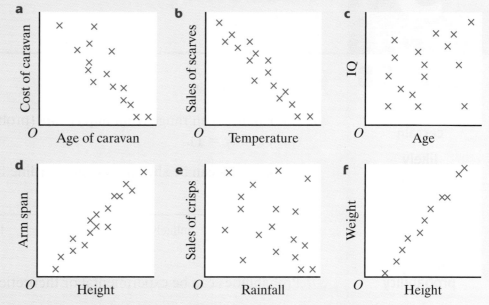

a
Cost of caravan / Age of caravan

b
Sales of scarves / Temperature

c
IQ / Age

d
Arm span / Height

e
Sales of crisps / Rainfall

f
Weight / Height

2 The table shows the age and second-hand value of cars.

Age of car (years)	3	1	4	10	6	9	8
Value of car (£)	3300	4800	2300	750	2600	300	1600

a Draw a scatter graph of the results.

b Describe the relationship shown by your scatter graph.

3 Annette collected the following information on the temperature and the number of visitors to a theme park.

Temperature (°C)	24	22	20	16	19	23	15	18	21	26
Number of visitors	720	480	440	240	510	550	280	500	600	700

a Draw a scatter graph and the line of best fit of the results.

b Use your line of best fit to estimate:

  i the number of people if the temperature is 17 °C

  ii the temperature if 150 people visit the theme park.

c Which of the above answers is most likely to be inaccurate?

  Give a reason for your answer.

5 Probability

Key words

- probability
- certain
- likely
- unlikely
- impossible
- outcome
- theoretical probability
- experimental probability
- mutually exclusive events
- sample space diagram

GET IT RIGHT!

Remember that probabilities can only take values between 0 and 1. Any other answers will need to be checked as they are WRONG.

Key points

◎ Probabilities can range from **impossible** (probability = 0) to **certain** (probability = 1).

Probabilities can be shown on a number line like this:

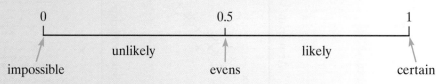

◎ Probabilities can be experimental or theoretical.

> Experimental probability is the probability arising from some experiment (sometimes called the relative frequency).

> Theoretical probability is based upon equally likely outcomes and gives an indication of what should happen in theory.

◎ Probabilities are usually expressed as fractions and the probability of an event is

$$\frac{\text{number of required outcomes}}{\text{total number of possible outcomes}}$$

◎ Events are **mutually exclusive** when they cannot happen at the same time.

The sum of the probabilities of all the mutually exclusive events is 1.

If event A and event B are mutually exclusive, then
P(A or B) = P(A) + P(B).

Worked Example Basic probability

Question

A bag contains 5 blue discs and 3 green discs.

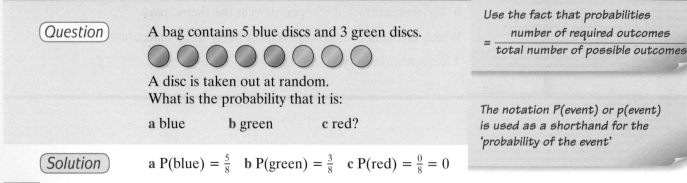

A disc is taken out at random.
What is the probability that it is:

a blue   **b** green   **c** red?

> Use the fact that probabilities
> $= \dfrac{\text{number of required outcomes}}{\text{total number of possible outcomes}}$

> The notation P(event) or p(event) is used as a shorthand for the 'probability of the event'

Solution

a P(blue) = $\frac{5}{8}$ **b** P(green) = $\frac{3}{8}$ **c** P(red) = $\frac{0}{8}$ = 0

Worked Example   Sample spaces

Question Two dice are thrown and their possible outcomes recorded on a sample space diagram.

a Complete a sample space diagram for the outcomes.

b Use the diagram to calculate:

    i the probability of throwing a double

    ii the probability of a total of 5.

Solution **a**

Second dice

First dice

	1	2	3	4	5	6
1	1,1	1,2	1,3	1,4	1,5	1,6
2	2,1	2,2	2,3	2,4	2,5	2,6
3	3,1	3,2	3,3	3,4	3,5	3,6
4	4,1	4,2	4,3	4,4	4,5	4,6
5	5,1	5,2	5,3	5,4	5,5	5,6
6	6,1	6,2	6,3	6,4	6,5	6,6

GET IT RIGHT!

It is useful to draw diagrams to help with these questions.

To work out probabilities for combined events, it is often useful to show all the possible outcomes in a table (called a sample space diagram).

a i P(throwing a double) $= \frac{6}{36} = \frac{1}{6}$ (the yellow cells)

   ii P(total of 5) $= \frac{4}{36} = \frac{1}{9}$ (the pink cells)

Remember to cancel down fractions where possible. Decimals and percentages can also be used to express probabilities.

Worked Example   Relative frequency

Question A bag contains 5 blue discs, 3 red discs and 2 white discs.

A disc is taken from the bag and the colour noted.

The term relative frequency is used to describe experimental probability.

It is then replaced.

a If the experiment is repeated 50 times, how many times would you expect a blue disc to be drawn?

The table shows the results obtained when the experiment is carried out 100 times.

Results from 100 draws			
	Blue	**Red**	**White**
Frequency	54	27	19

The expected number is found by multiplying the probability by the number of goes.

b i What is the relative frequency of a red disc?

   ii How does this compare with the theoretical probability?

EXAMINER SAYS...

Make sure you understand the difference between relative frequency (from experimental results) and theoretical probability (what should happen in theory).

Questions on relative frequency are generally answered badly in the examination.

Solution

a P(blue disc) $= \frac{5}{10}$ Expectation $= \frac{5}{10} \times 50 = 25$

b i Relative frequency of a red disc $= \frac{27}{100}$

   ii Expectation of a red disc $= \frac{3}{10} \times 100 = 30$

   The relative frequency is slightly lower.

   In this case, the relative frequency is $\frac{27}{100}$ whereas the theoretical probability $= \frac{3}{10} = \frac{30}{100}$

Worked Example Mutually exclusive events

(Question)

A card is drawn from a pack of 52 cards.

What is the probability that the chosen card is:

a a five

b a red card

c a Queen or a King

d a red card or a Jack?

> P(Queen or King) = P(Queen) + P(King) because the two events are mutually exclusive.

GET IT RIGHT!

Since the sum of the probabilities of all the mutually exclusive events is 1 then if the probability of something happening is p, the probability of it NOT happening is $1 - p$

a $P(\text{five}) = \frac{4}{52} = \frac{1}{13}$

b $P(\text{red card}) = \frac{26}{52} = \frac{1}{2}$

c $P(\text{Queen or King}) = P(\text{Queen}) + P(\text{King}) = \frac{4}{52} + \frac{4}{52} = \frac{8}{52} = \frac{2}{13}$

d $P(\text{red or Jack}) = \frac{28}{52} = \frac{7}{13}$

(Solution)

> $P(\text{red or Jack}) \neq P(\text{red}) + P(\text{Jack})$ because the two events are NOT mutually exclusive. In this case, the Jack of hearts and the Jack of diamonds would be counted twice.

A♥	2♥	3♥	4♥	5♥	6♥	7♥
8♥	9♥	10♥	J♥	Q♥	K♥	J♣
A♦	2♦	3♦	4♦	5♦	6♦	7♦
8♦	9♦	10♦	J♦	Q♦	K♦	J♠

Probability

END OF CHAPTER QUESTIONS

Time Yourself!

Can you complete these questions in **35** minutes?

1 Gordon has 30 marbles. Some are blue, some are red and some are green.

He picks a marble at random.

The probability that he will pick a blue marble is 0.3

The probability that he will pick a red marble is $\frac{1}{5}$

a How many blue marbles does Gordon have?

b How many red marbles does he have?

c What is the probability that he will pick a green marble?

2 The table shows the frequency distribution after drawing a card from a standard pack of 52 cards 60 times.

Results from 60 draws				
	club	heart	diamond	spade
Frequency distribution	11	17	10	22

a What is the relative frequency of getting a club?

b What is the relative frequency of getting a red card?

c What is the theoretical probability of getting a spade?

3 Two dice are thrown and their scores are added together.

Second dice

First dice		1	2	3	4	5	6
	1	2	3	4	5	6	
	2	3	4	5			
	3	4	5				
	4						
	5						
	6						

a Copy and complete the sample space diagram and use it to find

   i the probability of getting a total of 4

   ii the probability of getting a total of 7

   iii the probability of getting a total of 13.

b What is the modal score?

4 The probability that Jamie will win a swimming competition is 0.85

What is the probability that she will not win?

5 The table shows the probabilities that a student will choose a particular drink from a vending machine.

Cola	lemonade	orange	other
0.35	$\frac{1}{5}$		0.1

What is the probability that the student will choose orange?

1 The number of magazines sold each day by a village shop is recorded for one month.

Number of magazines	Frequency	
0	6	
1	7	
2	9	
3	4	
4	3	
5	1	
Total	30	

Calculate the mean number of magazines sold each day. *(3 marks)*

AQA Spec B, Foundation Module 1, Practice Paper, 2008

2 A police officer records the speeds of 60 cars on a dual carriageway.

Speed (mph)	Frequency	Midpoint	
40 to less than 50	9		
50 to less than 60	27		
60 to less than 70	21		
70 to less than 80	3		

Use the class midpoints to calculate an estimate of the mean speed of these cars. *(3 marks)*

3 Jane records the times taken by 30 students to complete a number puzzle.

Time t (minutes)	Number of students
$2 < t \le 4$	3
$4 < t \le 6$	6
$6 < t \le 8$	7
$8 < t \le 10$	8
$10 < t \le 12$	5
$12 < t \le 14$	1

Calculate an estimate of the mean time taken to complete the puzzle. *(3 marks)*

4 Beth records the time, in minutes, she spends doing homework each school night for three weeks.

    47   35   54   50   63

    38   49   72   66   62

    55   69   69   42   70

Beth attempts to draw an ordered stem-and-leaf diagram to show these amounts.

This is her diagram:

```
3 | 5  8
4 | 7  9  2
5 | 0  4  5
6 | 2  3  6  9
    0  2
```

Key 3|5 represents 35 minutes

Beth has made **three** mistakes in her diagram.

Write down the mistakes she has made.   *(3 marks)*

AQA Spec B, Higher Module 1, Practice Paper, 2008

5   The scatter graph shows the lengths, in centimetres (cm), and the weights, in kilograms (kg), of eight newborn babies.

   (a) Draw a line of best fit on the scatter graph.

                                 (1 mark)

   (b) Use your line of best fit to estimate the weight of a newborn baby whose length is 54 cm.   *(1 mark)*

Weights and lengths of babies

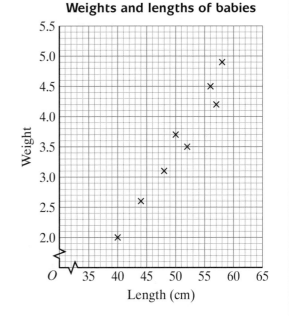

6   The letters of the word 'arithmetic' are written on cards.

A card is chosen at random.

The probabilities of the following events have been marked on the probability scale below.

T: The letter chosen is a T.

N: The letter chosen is **not** an I.

P: The letter chosen is a P.

Label each arrow with the letter to show which event it represents.   *(3 marks)*

AQA Spec B, Foundation Module 1, Practice Paper, 2008

7 A three-sided spinner has sections numbered 1, 2 and 3.

It is spun 20 times.

The results are shown below.

2	1	1	3	1	2	3	3	2	3
1	2	3	2	2	3	2	3	1	3

(a) Use these results to calculate the relative frequency of each number.

Number	1	2	3
Relative frequency			

(2 marks)

(b) The following table shows the relative frequency of each number after 100 spins.

Number	1	2	3
Relative frequency	0.14	0.45	0.41

Rafiq says that this table gives a better estimate for the probability of spinning the number 1.

Is he correct?

Give a reason for your answer.

(1 mark)

AQA Spec B, Foundation Module 1, Practice Paper, 2008

8 The graph shows the price index of the colour TV licence from 1995 to 2006.

The base year is 1995.

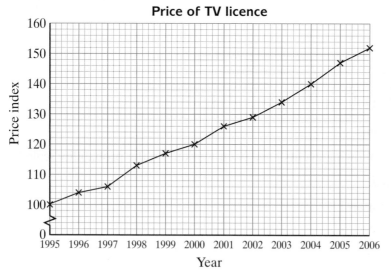

(a) In which year was the price of the licence 20% more than the price in 1995?   *(1 mark)*

(b) What was the percentage increase in the price of the licence from 1995 to 2006?   *(2 marks)*

AQA Spec B, Foundation Module 1, Nov 06

1   Sophie counts the number of letters in each word of the first sentence of a newspaper.

These are here results.

9   2   3   6   5   7   6   3   7   9   8   4   8   7

> There are 14 numbers so the median is $\frac{1}{2}$ the (14+1)th value ie 7.5th value so between the 7th and 8th values.

(a) Work out the median.

> The first mark is awarded for arranging the numbers in order of size.

2   3   3   4   5   6   6   7   7   7   8   8   9   9

> The candidate identifies that the middle number lies between 6 and 7

Answer   Median = 6.5   (2 marks)

> The second mark is awarded here for writing down the correct value for the median as 6.5 or $6\frac{1}{2}$

(b) Calculate the mean of these numbers.

$$\text{Mean} = \frac{9+2+3+6+5+7+6+3+7+9+8+4+8+7}{14}$$

$$= \frac{84}{14} = 6$$

> The first mark is awarded for the correct method of adding the numbers. The second mark is for dividing by 14 as there are 14 numbers altogether.

> It is a good idea to use the original data for adding up – just in case there is a mistake in reordering.

Answer   mean = 6   (3 marks)

> The final mark is awarded for the correct answer of 6

2   The number of minutes that a train arrived late at a station is shown in the table below.

> The candidate correctly finds the midpoints for the intervals for 1 mark.

Number of minutes late, t	Frequency	Midpoint	ft
$0 < t \le 10$	16	5	80
$10 < t \le 20$	10	15	150
$20 < t \le 30$	11	25	275
$30 < t \le 40$	8	35	280
$40 < t \le 50$	5	45	225

$\sum f = 50$ $\sum ft = 1010$

> It is useful to add extra columns for the 'ft' column and complete the totals to give $\sum f$ and $\sum ft$

(a) Complete the midpoint column and use it to calculate an estimate of the mean number of minutes that the trains arrive late.

> The method attracts a second mark making use of $\sum ft = 1010$ and $\sum f = 50$

$$\text{Mean} = \frac{\sum ft}{\sum f} = \frac{1010}{50} = 20.2$$

Answer   Mean = 20.2   (3 marks)

> The final mark is awarded for the correct answer of 20.2

(b) Which class interval contains the median number of minutes that trains arrived late?.

Number of minutes late, t	Frequency	
$0 < t \leq 10$	16	16 = 16
$10 < t \leq 20$	10	16 + 10 = 26
$20 < t \leq 30$	11	
$30 < t \leq 40$	8	
$40 < t \leq 50$	5	

The median is the middle value when the values are in order of size. Since there are 50 values then the median is the $\frac{(50+1)th}{2}$ = 25.5th value

The median is in the $10 < t \leq 20$ interval.

The 25.5th value is in the group, which contains the 25th and 26th values. The $10 < t \leq 20$ class interval contains the median

Answer $10 < t \leq 20$ minutes

(2 marks)

The two marks are awarded for identifying the middle, that is, the 25.5th value, and obtaining the interval from the table.

3 Phil wants to test if a six-sided dice is biased.

He rolls the dice 20 times.

Here are his results.

2 3 5 6 1 2 4 5 6 2 3 4 2 1 2 3 5 6 2 1

(a) Complete the relative frequency table.

	1	2	3	4	5	6
Tally	III	IIII I	III	II	III	III
Frequency	3	6	3	2	3	3

The candidate appreciates the difference between frequency and relative frequency.

It is helpful to cancel fractions where possible, but it is easier to compare fractions if they have the same denominator.

	1	2	3	4	5	6
Relative frequency	$\frac{3}{20}$	$\frac{6}{20} = \frac{3}{10}$	$\frac{3}{20}$	$\frac{2}{20} = \frac{1}{10}$	$\frac{3}{20}$	$\frac{3}{20}$

No marks are given for the tally chart – a method mark is awarded for calculating a relative frequency.

(b) Phil concludes that the dice is biased towards a number.

Write down the number that you think the dice is biased towards.

Explain your answer.

The answer is correct, but the explanation is also required in order to get the mark.

Number 2

Explanation The number 2 occurs much more frequently than the other numbers – they should all be about the same

(2 marks)

AQA Spec B, Foundation Module 1, Practice Paper, 2008

1 Integers and rounding

Key words

integer
less than (<)
greater than (>)
sum
product
factor
multiple
least common multiple (LCM)
common factor
highest common factor (HCF)
prime number
index notation
reciprocal
significant figures
decimal places
estimate
upper bound
lower bound

Key points

- An **integer** is any positive or negative **whole** number or zero.

- A **factor** is a natural number that divides exactly into a number (no remainder). The **highest common factor (HCF)** of two (or more) numbers is the highest number that divides exactly into both (all) of them.

- **Multiples** of a number are the products of its multiplication table. The **least common multiple (LCM)** is the lowest multiple that is common to two or more numbers.

- A **prime number** is a number with exactly two factors, itself and 1.

- To find the **sum** of numbers, **add** them.

- To find the **difference** between two numbers, **subtract** them.

- To find the **product** of numbers, **multiply** them.

- **Index notation** The number 4 in 2^4 is the index (plural **indices**).

- 1 divided by a number gives its **reciprocal**, for example, the reciprocal of 5 is $\frac{1}{5}$

  Any number multiplied by its **reciprocal** equals 1.

- The **upper bound** is the maximum possible value of a measurement.

  The **lower bound** is the minimum possible value of a measurement.

Worked Example Factors, multiples and directed numbers

Question

Here is a list of numbers: 6   9   10   13   24   27

From this list, write down the number that is:

a a multiple of 12   **b** a factor of 12   **c** a prime number

d the reciprocal of $\frac{1}{9}$ **e** -3×-9 **f** $-2 - -8$

Solution

Because $2 \times 12 = 24$

a 24 is a multiple of 12   **b** 6 is a factor of 12   Because 6 divides exactly into 12

Because 13 has just two factors (1 and 13)

c 13 is a prime number   **d** the reciprocal of $\frac{1}{9}$ is $\frac{9}{1} = 9$

BUMP UP C THE GRADE

To get a grade **C** you must be able to recognise prime numbers and find the reciprocal of a number.

e $-3 \times -9 = 27$

f $-2 - -8 = -2 + 8 = 6$
Start at -2 and go up by 8

When multiplying or dividing:

signs same = positive

signs different = negative

Worked Example Highest common factor and least common multiple

(Question)

a Find the highest common factor (HCF) of 18 and 24.

b Find the least common multiple (LCM) of 18 and 24.

(Solution)

a The factors of 18 are   1   2   3   6   9   18

The factors of 24 are   1   2   3   4   6   8   12   24

The HCF of 18 and 24 is 6.

1, 2, 3 and 6 are all **common factors** of 18 and 24

6 is the **highest** number that is **common** to both lists

b The multiples of 18 are   18   36   54   72   90...

The multiples of 24 are   24   48   72   96...

The LCM of 18 and 24 is 72.

72 is the **lowest** number that is **common** to both lists

BUMP UP

C

THE GRADE To get a grade **C** you must be able to find LCMs and HCFs.

Worked Example Writing a number as a product of its prime factors

(Question)

a Write 360 as a product of its prime factors in index form.

b Write 3600 as a product of its prime factors in index form.

(Solution)

a

2	360
2	180
2	90
3	45
3	15
5	5
	1

Try the prime numbers in order to see if they are factors.

360 written as a product of its prime factors is $2 \times 2 \times 2 \times 3 \times 3 \times 5$

In index form $360 = 2^3 \times 3^2 \times 5$

The index in 2^3 tells you that 3 twos are multiplied together.

b $3600 = 360 \times 10 = 360 \times 2 \times 5$

The extra 2 and 5 increase the indices of 2 and 5 by 1

So $3600 = 2^4 \times 3^2 \times 5^2$

BUMP UP

C

THE GRADE To get a grade **C** you must be able to write a number as a product of its prime factors.

Worked Example Estimating

Question Use approximations to estimate the value of: **a** $\dfrac{49.1 \times 9.08}{2.87}$ **b** $\dfrac{972 + 215}{8.09 - 3.78}$

Solution

a $\dfrac{49.1 \times 9.08}{2.87} \approx \dfrac{50 \times 9}{3} = \dfrac{450}{3} = 150$

This means round all numbers to 1 significant figure before doing the calculation.

b $\dfrac{972 + 215}{8.09 - 3.78} \approx \dfrac{1000 + 200}{8 - 4} = \dfrac{1200}{4} = 300$

AQA EXAMINER SAYS...

Candidates sometimes get no marks because they do an accurate calculation instead of an estimate.

AQA EXAMINER SAYS...

You might be asked to do a calculation like part **b** accurately. If so, take care – you need to do the numerator and denominator separately first *or* use brackets, that is, (972 + 215) ÷ (8.09 – 3.78) gives the correct answer.

GET IT RIGHT!

First work out sums/differences in the numerator/denominator.

BUMP UP THE GRADE C To get a grade **C** you must be able to divide by a decimal, for example, if the denominator in part **a** was 0.3, multiplying top and bottom by 10 would give $\dfrac{450}{0.3} = \dfrac{4500}{3} = 1500$

Worked Example Maximum and minimum values

Question A plan shows the length of a field is 42 metres to the nearest metre.

What are the upper and lower bounds of the length of the field?

Solution

The upper bound is 42.5 m. ◄—— The upper bound is (42 + 0.5) m

The lower bound is 41.5 m. ◄—— The lower bound is (42 − 0.5) m

This can be written as 41.5 m ≤ length of field < 42.5 m

≤ means 'less than or equal to'

41.5 m is the minimum possible length.

< means 'less than'.

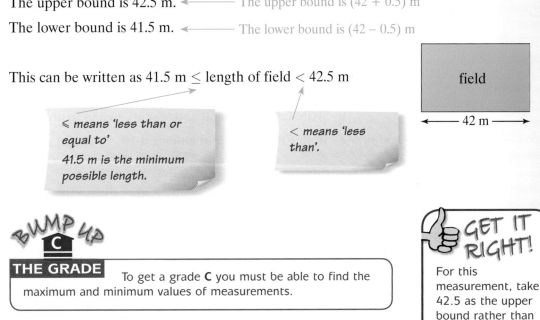

field

←—— 42 m ——→

AQA EXAMINER SAYS...

Take care with upper bounds. For example, if the number of students in a school is 700 to the nearest 10, the upper bound is 704 (whole numbers); if the weight of a cabbage is 700g to the nearest 10g, the upper bound is 705g (measurements).

BUMP UP THE GRADE C To get a grade **C** you must be able to find the maximum and minimum values of measurements.

GET IT RIGHT!

For this measurement, take 42.5 as the upper bound rather than 42.4999...

Integers and rounding

1 Work out   **a** $\dfrac{6-9}{3}$ **b** $\dfrac{-1+9}{-2}$ **c** $\dfrac{-5 \times 8}{-4}$ **d** $\dfrac{-7 \times -6}{4}$

2 Here is a list of numbers.   15   26   35   37   40   54   60   72

From this list, write down:

a two numbers which have a sum of 50

b two numbers which have a difference of 32

c the number which is a factor of 30

d the number which is a multiple of 7

e the number which is the product of 6 and 9

f a prime number

3 Find all the common factors of: **a** 24 and 32   **b** 36 and 48

4 Find the highest common factor of:

a 20 and 36   **b** 30 and 45   **c** 48 and 72

5 Find the least common multiple of:

a 6 and 15   **b** 12 and 20   **c** 20 and 36

6 Write each number as the product of its prime factors in index notation.

a 28   **b** 54   **c** 90   **d** 120   **e** 250   **f** 567

7 Find the reciprocal of: **a** 8   **b** $\frac{1}{3}$ **c** 0.4   **d** $\frac{5}{8}$ **e** −4

8 a Write eight million in figures.

b Write eight thousand and seventy-two in figures.

c Write 7829   **i** to the nearest 10   **ii** to 1 significant figure

9 Use approximations to estimate the value of the following calculations.

You **must** show your working.

a $\dfrac{206 \times 4.91}{1.92^2}$ **b** $\dfrac{5.74 + 3.12}{0.879}$ **c** $\dfrac{83.9}{19.3 \times 0.163}$ **d** $\dfrac{86.71 - 32.04}{41.56 + 79.52}$

10 The capacity of a paddling pool is 800 litres to the nearest 10 litres.

What are the upper and lower bounds of the capacity?

Decimals and fractions

Key words

digit

numerator

denominator

terminating decimal

recurring decimal

unit fraction

proper fraction

improper fraction or top-heavy fraction

mixed number or mixed fraction

equivalent fraction

simplify

Make sure you can also do all the decimal and fraction calculations on a calculator.

Key points

- The place values of **decimal digits** are fractions, for example, in 0.37, the value of the 3 is $\frac{3}{10}$ and the value of the 7 is $\frac{7}{100}$

  To change a decimal to a fraction, use the last place value in the denominator, for example, $0.379 = \frac{379}{1000}$

  To change a fraction to a decimal, divide the numerator by the denominator.

- A **recurring** decimal has a repeating digit or group of digits, for example, 0.444… written as $0.\dot{4}$ and 0.7305305305… written as $0.7\dot{3}0\dot{5}$

- To **add** or **subtract** decimals, **line up the decimal points**.

- When a decimal is **multiplied or divided by 10**, the digits move **1 place**, when **multiplied or divided by 100**, the digits move **2 places** etc., for example, $12.76 \times 10 = 127.6$, $34.8 \times 100 = 3480$, $0.209 \times 1000 = 209$
  $12.76 \div 10 = 1.276$, $34.8 \div 100 = 0.348$, $0.209 \div 1000 = 0.000\,209$

- **To multiply decimals**, remove the decimal points, multiply as usual, then count the number of decimal places and insert the decimal point accordingly.

- **To divide a decimal**, write as a fraction, multiply numerator and denominator by 10 until the denominator becomes a whole number, then divide as usual.

- **To simplify a fraction**, divide the numerator and denominator by the same numbers until you find the simplest equivalent form.

- **To put fractions in order** or **add** or **subtract fractions**, write them with the same denominator.

- **To find a fraction of something**, divide it by the denominator of the fraction, then multiply by the numerator.

- **To multiply fractions**, multiply the numerators and multiply the denominators. (Change mixed numbers to improper fractions first.)

- **To divide by a fraction**, multiply by its reciprocal (upside-down) fraction.

- The **reciprocal** of $\frac{3}{2}$ is $\frac{2}{3}$. The reciprocal of 0.2 is $\frac{1}{0.2} = \frac{10}{2} = 5$

BUMP UP

C

THE GRADE To get a grade **C** you must be able to find the reciprocal of a decimal.

Worked Example Putting decimals in order

Question

Put these decimals in order of size, starting with the smallest.

1.365   1.4   1.192   1.28

Write them all with 3 decimal places:

1.365, 1.400, 1.192, 1.280

Solution

The correct order is 1.192   1.28   1.365   1.4

Worked Example Decimal calculations

Question

a 1.7 + 0.84   **b** 3.4 − 1.52   **c** 0.3 × 2.5 + 4.5 ÷ 100   **d** 2.52 ÷ 0.6

Solution

a 1.7 + 0.84 = 2.54

$$\begin{array}{r} 1.70 \\ +0.84 \\ \hline 2.54 \\ \scriptstyle 1 \end{array}$$

b 3.4 − 1.52 = 1.88

$$\begin{array}{r} {}^2\!3\,.\,{}^{13}\!4\,{}^1\!0 \\ -1\,.\,5\,2 \\ \hline 1\,.\,8\,8 \end{array}$$

To add or subtract, line up the decimal points.

Put a 0 in the space.

👍 **GET IT RIGHT!**

Remember the order of operations:
Brackets
Indices
Divide
Multiply
Add
Subtract

c 0.3 × 2.5 + 4.5 ÷ 100

= 0.75   +   0.045 = 0.795

d 2.52 ÷ 0.6 = $\dfrac{2.52}{0.6} = \dfrac{25.2}{6} = 4.2$

×10

$$\begin{array}{r} 4\,.\,2 \\ 6\,\overline{)2\,5\,.\,{}^1\!2} \end{array}$$

×10

× and ÷ before +

To multiply, ignore decimal points. After multiplying, count the number of decimal places in the calculation (2 here) – the answer has the same number.

Multiply top and bottom by 10 until the bottom is a whole number. The answer you get will be the same as for the original numbers.

AQA EXAMINER SAYS...

Candidates often lose marks because they don't know how to multiply and divide decimals.

BUMP UP

C

THE GRADE   To get a grade **C** you must be able to divide by a decimal.

Worked Example Putting fractions in order

Question

Arrange these fractions in order of size. Start with the smallest.

$$\frac{2}{3} \quad \frac{5}{8} \quad \frac{7}{12} \quad \frac{5}{6} \quad \frac{3}{4}$$

Change each fraction to an equivalent fraction with denominator 24

Solution

All the denominators are factors of 24.

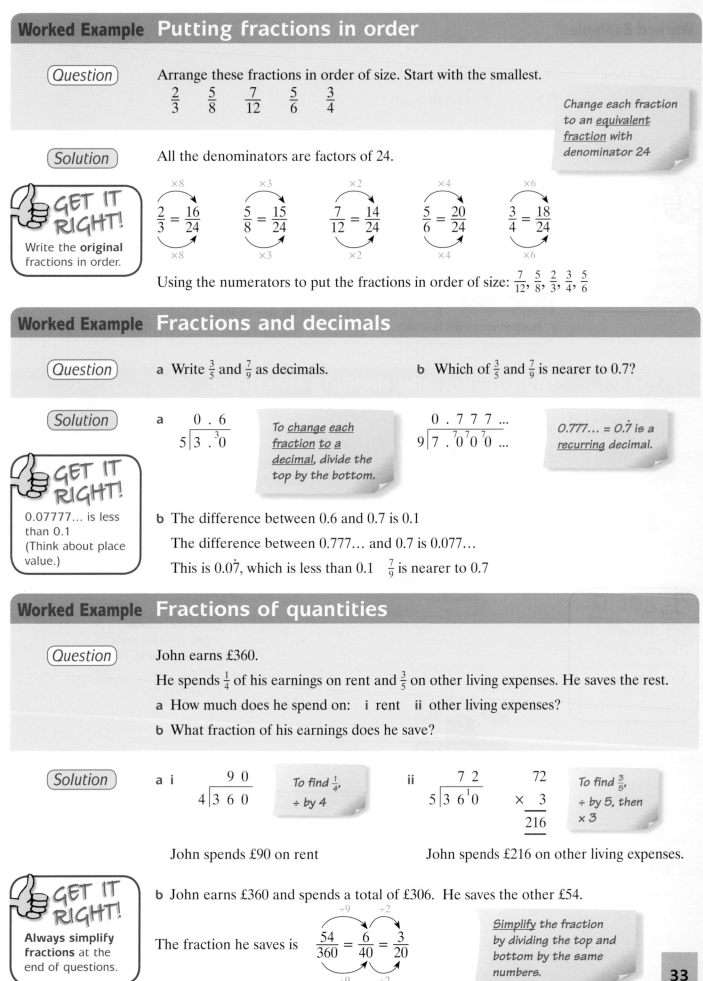

GET IT RIGHT!

Write the **original** fractions in order.

$$\frac{2}{3} = \frac{16}{24} \quad \frac{5}{8} = \frac{15}{24} \quad \frac{7}{12} = \frac{14}{24} \quad \frac{5}{6} = \frac{20}{24} \quad \frac{3}{4} = \frac{18}{24}$$

Using the numerators to put the fractions in order of size: $\frac{7}{12}, \frac{5}{8}, \frac{2}{3}, \frac{3}{4}, \frac{5}{6}$

Worked Example Fractions and decimals

Question

a Write $\frac{3}{5}$ and $\frac{7}{9}$ as decimals.

b Which of $\frac{3}{5}$ and $\frac{7}{9}$ is nearer to 0.7?

Solution

GET IT RIGHT!

0.07777... is less than 0.1
(Think about place value.)

a

$$\begin{array}{r} 0\;.\;6 \\ 5\overline{\smash)3\;.\;{}^3 0} \end{array}$$

To change each fraction to a decimal, divide the top by the bottom.

$$\begin{array}{r} 0\;.\;7\;7\;7\;... \\ 9\overline{\smash)7\;.\;{}^7 0 {}^7 0 {}^7 0 ...} \end{array}$$

0.777... = 0.$\dot{7}$ is a recurring decimal.

b The difference between 0.6 and 0.7 is 0.1

The difference between 0.777... and 0.7 is 0.077...

This is 0.0$\dot{7}$, which is less than 0.1 $\frac{7}{9}$ is nearer to 0.7

Worked Example Fractions of quantities

Question

John earns £360.

He spends $\frac{1}{4}$ of his earnings on rent and $\frac{3}{5}$ on other living expenses. He saves the rest.

a How much does he spend on: i rent   ii other living expenses?

b What fraction of his earnings does he save?

Solution

a i

$$\begin{array}{r} 9\;0 \\ 4\overline{\smash)3\;6\;0} \end{array}$$

To find $\frac{1}{4}$, ÷ by 4

ii

$$\begin{array}{r} 7\;2 \\ 5\overline{\smash)3\;6^1 0} \end{array} \qquad \begin{array}{r} 72 \\ \times\;\;3 \\ \hline 216 \end{array}$$

To find $\frac{3}{5}$, ÷ by 5, then × 3

John spends £90 on rent   John spends £216 on other living expenses.

GET IT RIGHT!

Always simplify fractions at the end of questions.

b John earns £360 and spends a total of £306. He saves the other £54.

The fraction he saves is $\dfrac{54}{360} = \dfrac{6}{40} = \dfrac{3}{20}$

Simplify the fraction by dividing the top and bottom by the same numbers.

Worked Example — Adding and subtracting fractions

Question

Helen has $5\frac{1}{3}$ metres of fabric. She uses $2\frac{1}{2}$ metres to make a skirt.

What length of fabric is left?

Solution

$$5\frac{1}{3} - 2\frac{1}{2} = 3 + \frac{1}{3} - \frac{1}{2}$$

$$= 3 + \frac{2}{6} - \frac{3}{6} \quad \longleftarrow \text{Change both fractions to sixths.}$$

$$= 2 + \frac{8}{6} - \frac{3}{6} \quad \longleftarrow \text{Change 1 unit to } \frac{6}{6} \text{ and do the subtraction.}$$

There are $2\frac{5}{6}$ metres of fabric left.

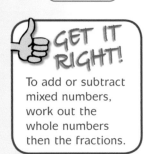

GET IT RIGHT!

To add or subtract mixed numbers, work out the whole numbers then the fractions.

THE GRADE To get a grade **C** you must be able to solve fraction problems in words.

Worked Example — Multiplying and dividing fractions

Question

Work out:

a $\dfrac{3}{4} \times \dfrac{2}{5}$

b $\dfrac{9}{16} \div \dfrac{3}{8}$

Solution

a $\dfrac{3}{\cancel{4}_2} \times \dfrac{\cancel{2}^1}{5}$

$= \dfrac{3}{10}$

Simplify before multiplying where possible.

b $\dfrac{9}{16} \div \dfrac{3}{8} = \dfrac{\cancel{9}^3}{\cancel{16}_2} \times \dfrac{\cancel{8}^1}{\cancel{3}_1}$

$= \dfrac{3}{2} = 1\frac{1}{2}$

Just the second fraction is turned upside-down.

Change improper fractions to mixed numbers at the end.

GET IT RIGHT!

To cancel one number must be on the top and one on the bottom

THE GRADE To get a grade **C** you must be able to divide fractions.

Decimals and fractions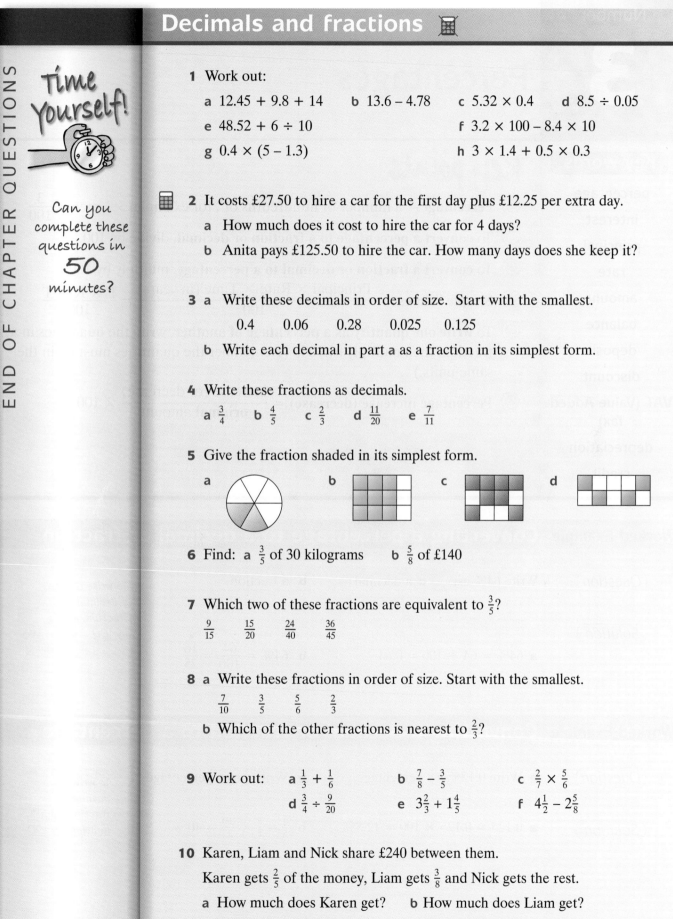

1 Work out:

   a $12.45 + 9.8 + 14$ **b** $13.6 - 4.78$ **c** 5.32×0.4 **d** $8.5 \div 0.05$

   e $48.52 + 6 \div 10$ **f** $3.2 \times 100 - 8.4 \times 10$

   g $0.4 \times (5 - 1.3)$ **h** $3 \times 1.4 + 0.5 \times 0.3$

2 It costs £27.50 to hire a car for the first day plus £12.25 per extra day.

   a How much does it cost to hire the car for 4 days?

   b Anita pays £125.50 to hire the car. How many days does she keep it?

3 a Write these decimals in order of size. Start with the smallest.

       0.4   0.06   0.28   0.025   0.125

   b Write each decimal in part **a** as a fraction in its simplest form.

4 Write these fractions as decimals.

   a $\frac{3}{4}$ **b** $\frac{4}{5}$ **c** $\frac{2}{3}$ **d** $\frac{11}{20}$ **e** $\frac{7}{11}$

5 Give the fraction shaded in its simplest form.

   a   **b**   **c**   **d**

6 Find: **a** $\frac{3}{5}$ of 30 kilograms   **b** $\frac{5}{8}$ of £140

7 Which two of these fractions are equivalent to $\frac{3}{5}$?

   $\frac{9}{15}$ $\frac{15}{20}$ $\frac{24}{40}$ $\frac{36}{45}$

8 a Write these fractions in order of size. Start with the smallest.

     $\frac{7}{10}$ $\frac{3}{5}$ $\frac{5}{6}$ $\frac{2}{3}$

   b Which of the other fractions is nearest to $\frac{2}{3}$?

9 Work out: **a** $\frac{1}{3} + \frac{1}{6}$ **b** $\frac{7}{8} - \frac{3}{5}$ **c** $\frac{2}{7} \times \frac{5}{6}$

                d $\frac{3}{4} \div \frac{9}{20}$ **e** $3\frac{2}{3} + 1\frac{4}{5}$ **f** $4\frac{1}{2} - 2\frac{5}{8}$

10 Karen, Liam and Nick share £240 between them.

    Karen gets $\frac{2}{5}$ of the money, Liam gets $\frac{3}{8}$ and Nick gets the rest.

   a How much does Karen get?   **b** How much does Liam get?

   c How much does Nick get?   **d** What fraction does Nick get?

Key words

percentage

interest

principal

rate

amount

balance

deposit

discount

VAT (Value Added Tax)

depreciation

credit

Key points

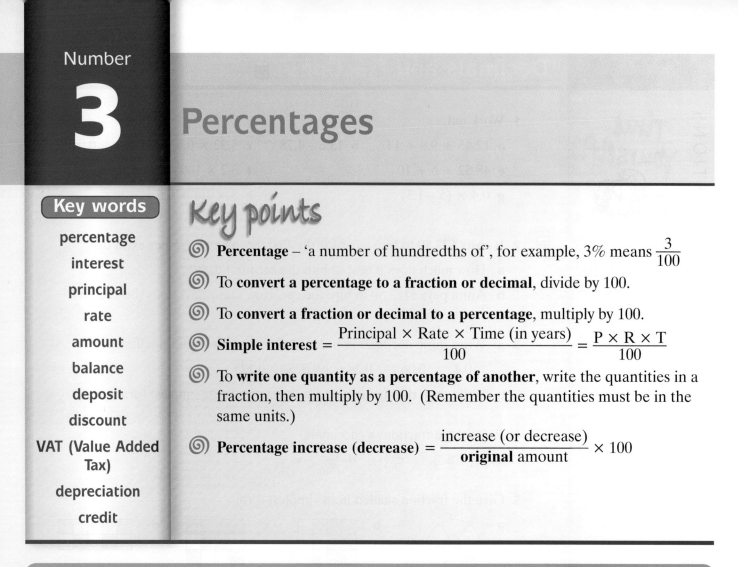

- ◉ **Percentage** – 'a number of hundredths of', for example, 3% means $\frac{3}{100}$

- ◉ To **convert a percentage to a fraction or decimal**, divide by 100.

- ◉ To **convert a fraction or decimal to a percentage**, multiply by 100.

- ◉ **Simple interest** $= \dfrac{\text{Principal} \times \text{Rate} \times \text{Time (in years)}}{100} = \dfrac{P \times R \times T}{100}$

- ◉ To **write one quantity as a percentage of another**, write the quantities in a fraction, then multiply by 100. (Remember the quantities must be in the same units.)

- ◉ **Percentage increase (decrease)** $= \dfrac{\text{increase (or decrease)}}{\text{original amount}} \times 100$

Worked Example Converting a percentage to a decimal or fraction

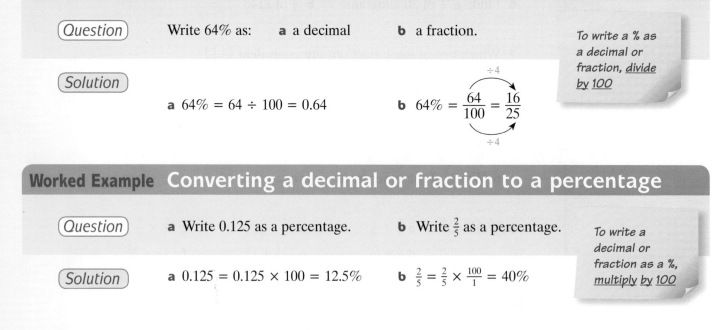

Question Write 64% as: **a** a decimal   **b** a fraction.

> To write a % as a decimal or fraction, *divide by 100*

Solution

a $64\% = 64 \div 100 = 0.64$

b $64\% = \dfrac{64}{100} = \dfrac{16}{25}$
($\div 4$ top and bottom)

Worked Example Converting a decimal or fraction to a percentage

Question **a** Write 0.125 as a percentage.   **b** Write $\frac{2}{5}$ as a percentage.

> To write a decimal or fraction as a %, *multiply by 100*

Solution **a** $0.125 = 0.125 \times 100 = 12.5\%$ **b** $\frac{2}{5} = \frac{2}{5} \times \frac{100}{1} = 40\%$

Worked Example — Working out a percentage of a quantity

Question   Find 45% of £12.80

Solution

Without a calculator	With a calculator
10% = £12.80 ÷ 10 *÷ 10 to find 10%* = £1.28 5% = $\frac{1}{2}$ of 10% = £0.64 (or 64p) 10% £1.28 + 10% + £1.28 + 10% + £1.28 + 10% + £1.28 + 5% + £0.64 ‾‾‾‾‾‾‾‾‾‾‾‾‾ 45% = £5.76 *Alternatively:* *£1.28 × 4* *= £5.12* *£5.12 + £0.64* *= £5.76*	1% = £12.80 ÷ 100 45% = £12.80 ÷ 100 × 45 = £5.76 *To find a percentage of an amount, ÷ by 100 to find 1%, then × by the percentage you need.* **Alternative method:** 45% = $\frac{45}{100}$ = 0.45 0.45 × £12.80 = £5.76

Worked Example — Increasing by a percentage

Question   A builder charges £240 plus VAT at $17\frac{1}{2}$ % for laying a path. What is the total bill?

Solution

👍 GET IT RIGHT!

Give the total bill, not just the VAT.

Without a calculator	With a calculator
10% = £240 ÷ 10 = £24 5% = $\frac{1}{2}$ of 10% = £12 2.5% = $\frac{1}{2}$ of 5% = £6 ‾‾‾‾‾‾‾‾‾‾‾‾‾‾‾‾ 17.5% (VAT) = £42 *17.5%* *= 10%* *+ 5%* *+ 2.5%* Total bill = £240 + £42 = £282	17.5% = £240 ÷ 100 × 17.5 = £42 (or 0.175 × £240 = £42) Total bill = £42 + £240 = £282 **Alternative method:** Total bill = 1.175 × £240 = £282 *100% + 17.5% = 117.5% = 1.175*

Worked Example — Decreasing by a percentage

Question   A new car costs £18 000. Its value falls by 20% in a year.

What is the value of the car at the end of the year?

Solution

GET IT RIGHT!

Give the final value, not the reduction.

Without a calculator	With a calculator
10% $= £18\,000 \div 10 = £1800$ $+ \underline{10\%} = +£1800$ $20\% = £3600$ Value after 1st year $= £18\,000 -$ Alternatively: $\qquad\qquad £3\,600$ $80\% = 10\% \times 8$ $\qquad\quad \underline{£14\,400}$ $= £1800 \times 8 = £14\,400$	$20\% = £18\,000 \div 100 \times 20 = £3600$ (or $0.2 \times £18\,000 = £3600$) Value after 1st year $= \qquad £18\,000 -$ Alternatively: $\qquad\qquad\qquad £3\,600$ $100\% - 20\% = 80\% = 0.8 \qquad \underline{£14\,400}$ $0.8 \times £18\,000 = £14\,400$

Worked Example ## Expressing one quantity as a percentage of another

Question

30 centimetres of a 5-metre roll of ribbon is damaged. What percentage is this?

Solution

GET IT RIGHT!

Both quantities must be in the **same units**.

$5 \text{ m} = 500 \text{ cm}$

$\dfrac{30}{500} = \dfrac{30}{{}_5\cancel{500}} \times \cancel{100} = 6\%$

Write as a fraction, then multiply by 100 to change to a percentage.

On a calculator:
$30 \div 500 = 0.06 = 6\%$

Worked Example ## Writing a decrease as a percentage

Question

When Peter goes on a diet, his weight falls from 75 kg to 72.5 kg.

What is the percentage decrease in Peter's weight?

Solution

Decrease in weight $= 75 - 72.5 = 2.5 \text{ kg}$

On a calculator:
$2.5 \div 75 =$
$0.0333... = 3.\dot{3}\%$

Percentage decrease $= \dfrac{2.5}{{}_3\cancel{75}} \times \cancel{100}^{4} = \dfrac{10}{3} = 3\tfrac{1}{3}\%$

Worked Example ## Writing an increase as a percentage

Question

A shopkeeper buys computer games for £15 each and sells them for £19.95 each.

What is the percentage profit?

On a calculator: $4.95 \div 15$
$= 0.33 = 33\%$

Solution

GET IT RIGHT!

The denominator must be the **original** amount

Profit $= £19.95 - £15 = £4.95$

Working in pounds:

Percentage profit $= \dfrac{4.95}{15} \times 100 = \dfrac{\cancel{495}^{99}}{\cancel{15}_3}$
$\qquad\qquad\qquad = 33\%$

Working in pence:

Percentage profit $= \dfrac{495}{1500} \times 100 = \dfrac{\cancel{495}^{99}}{\cancel{15}_3}$
$\qquad\qquad\qquad = 33\%$

Percentages

Time Yourself!

Can you complete these questions in

45

minutes?

END OF CHAPTER QUESTIONS

1 Write these percentages as fractions:

   a 25%   **b** 60%   **c** 45%   **d** 32%

2 Write these percentages as decimals:

   a 75%   **b** 7%   **c** 30%   **d** 55%

3 Write these as percentages:

   a 0.03   **b** 0.7   **c** 0.42   **d** 0.025

   e $\frac{3}{4}$ **f** $\frac{9}{10}$ **g** $\frac{4}{5}$ **h** $\frac{13}{20}$ **i** $\frac{7}{8}$

4 What percentage of each shape is shaded?

   a   **b**   **c**

5 Write 0.7, $\frac{5}{8}$ and 65% in order of size. Start with the smallest.

6 Find: **a** 40% of 650 metres   **b** 35% of £64   **c** 75% of 156 kilograms

7 A late offer gives 25% discount on a holiday that usually costs £580.

   What does the holiday cost after the discount?

8 Find the price of these goods after VAT is added at 17.5%.

 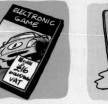 

   a   **b**   **c**

> Make sure you can also do all the questions with a calculator.

9 Andy earns £360 per week. He is awarded a pay rise of 4.5%.

   How much does he earn each week after the pay rise?

10 a A candidate gets 56 out of 80 marks in an exam. What percentage is this?

    b 15 candidates out of 60 who sit the exam fail. What percentage pass?

11 Dylan bought a mountain bike for £480 and sold it for £320.

   What was the percentage loss?

12 A shopkeeper buys a box of 10 pens for £4 and sells them for 65 pence each.

   What is the percentage profit?

Ratio and proportion

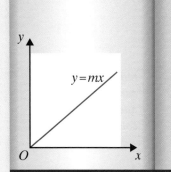

Key points

◎ A **ratio** compares the sizes of two or more quantities or numbers.
A **proportion** compares one part with the whole, for example,
if 11 boys and 13 girls go on a trip, the ratio of boys to girls is $11:13$.
The **proportion** of boys on the trip is $\frac{11}{24}$ and the **proportion** of girls is $\frac{13}{24}$.

◎ To **simplify a ratio**, divide both (all) numbers by the same number or
multiply them by the same number, for example, if 4 teachers and 24
students go on a school trip, the ratio of the number of teachers to the
number of students is $4:24 = 1:6$ (dividing by 4).

◎ **Unit ratio** – a ratio in the form $1:n$ or $n:1$, for example, $1:5$ or $3.75:1$

◎ **Unitary method** – a way of calculating quantities that are in proportion,
for example, if a recipe for 10 mince pies requires 250 g of mincemeat, but
you want to make 16 mince pies, you can first find the amount needed for 1
mince pie:
10 mince pies need 250 g of mincemeat
1 mince pie needs $\frac{250}{10}$ g = 25 g
16 mince pies need $25 \times 16 = 400$ g of mincemeat

Alternatively:
Amount for 6 mince
pies = $25 \times 6 = 150$ g
and add it to 250 g.

◎ **Direct proportion** If two variables are in direct proportion, one is equal to
a **constant** multiple of the other. The relationship can be written as $y = mx$
where x and y are variables and m is a **constant**. The graph showing this
relationship is a **straight line through the origin, (0, 0)**.

Worked Example **Finding and simplifying ratios**

Question
A school has 64 teachers and 800 students.

Write the ratio of students to teachers in: **a** its simplest form **b** the form $n:1$

Solution

Make sure you
write the quantities
in the **right order**.

a Ratio of students: teachers $= 800:64$

This is simplified by dividing $= 100:8$
both numbers by 8, then 4
$= 25:2$

b Ratio of students: teachers $= 25:2$

You could also get this by $= 12.5:1$
dividing 800:64 by 64

Simplest form
means using the
smallest possible
whole numbers.

To get 1 in the 2nd
part of the ratio, you
must divide by 2

EXAMINER SAYS...

Always write ratios
in the form $a : b$
and remember
that the order is
important.

Worked Example — Using the same units in ratios

Question

In a race, competitors swim 750 metres, cycle 20 kilometres and run 5 kilometres.

Write the swimming, cycling and running distances as a ratio in its simplest form.

Solution

Write the parts in the same units first.

1 km = 1000 m

Swimming distance : cycling distance : running distance

$= 750 \text{ m} : 20 \text{ km} : 5 \text{ km}$

$= 750 \text{ m} : 20\,000 \text{ m} : 5000 \text{ m}$

$= 750 : 20\,000 : 5000$

(Dividing each number by 10) $= 75 : 2000 : 500$

(Dividing each number by 25 or by 5, then 5 again) $= 3 : 80 : 20$

Worked Example — Using ratios to find quantities

Question

A factory employs 480 people. The ratio of men to women is 3 : 5

How many women are employed?

Solution

Make sure you give the right part of the ratio.

There are 3 men to every 5 women.

This means $\frac{3}{8}$ of the workforce are men and $\frac{5}{8}$ are women.

The number of women is $\frac{5}{8}$ of 480

There are 300 women.

$$\frac{60}{8\overline{)480}}$$

$$\begin{array}{r} 60 \\ \times\ 5 \\ \hline 300 \end{array}$$

Divide the people into 8 equal 'parts'. 5 of these parts are women.

Worked Example — Working with more difficult ratios

Question

When Ann, Bob and Carol set up a business Ann invested £2000, Bob invested £3000 and Carol invested £5000. In the first year the business makes a profit of £1600.

Divide the profit between Ann, Bob and Carol so their shares are in the same ratio as the amounts they invested.

Solution

Take care with the order so that you know which part is which.

Ann's share : Bob's share : Carol's share $= 2000 : 3000 : 5000$

$= 2 : 3 : 5$ ◄——— Simplify the ratio first

This means Ann gets $\frac{2}{10}$ of the profit, Bob gets $\frac{3}{10}$ and Carol gets $\frac{5}{10}$ ◄—— $2 + 3 + 5 = 10$

1 part $= \frac{1}{10}$ of £1600 = £160 ◄——— The profit is split into 10 equal parts. Ann gets 2, Bob gets 3 and Carol gets 5

Ann's share $= £160 \times 2 = £320$

Bob's share $= £160 \times 3 = £480$

Carol's share $= £160 \times 5 = £800$ Check that the total is £1600

BUMP UP

C

THE GRADE To get a grade **C** you must be able to solve more complex ratio problems like the above example.

Worked Example — Using ratios in different ways

Question

The recipe for a drink says 'Mix 2 parts pineapple juice with 3 parts orange juice'.

a How much pineapple juice should you mix with 600 ml of orange juice?

b How much pineapple juice do you need to make 1.5 litres of the drink?

Solution

Make sure you use the right parts of the ratio.

a 3 parts orange juice = 600 ml, so 1 part = 200 ml

2 parts pineapple juice = 400 ml

There are 600 ml of orange juice and the recipe says this is 3 parts. Find 1 part, then the quantity needed.

b 5 parts = 1.5 litres = 1500 ml ←———— 1.5 litres is the total amount, so this is 5 parts

1 part = 300 ml   2 parts pineapple juice = 600 ml

Worked Example — Increasing in proportion (using the unitary method)

Question

Seven pens cost £5.95. How much does it cost for nine of the same pens?

Solution

7 pens cost £5.95

1 pen costs 85 pence

9 pens cost £7.65

$$7\overline{)5\ 9^3 5}$$ giving 8 5

$$\begin{array}{r} 8\ 5 \\ \times\ \ 9 \\ \hline 7\ 6_4 5 \end{array}$$

Divide by 7 to find the cost of 1 pen. Then multiply by 9 to find the cost of 9 pens.

Worked Example — Decreasing in proportion

Question

A recipe for 24 biscuits uses 150 grams of currants.

What weight of currants do you need to make 16 biscuits?

Solution

24 biscuits need 150 g of currants

8 biscuits need 50 g of currants

16 biscuits need 100 g of currants.

÷ 3 then × 2 is much easier than the unitary method here.

150 ÷ 24 × 16 gives the same answer but is much more difficult.

Look out for quick ways like this.

BUMP UP

THE GRADE C   To get a grade **C** you must be able to solve proportion problems using the unitary or other methods.

Ratio and proportion

1 Simplify the following ratios:

  a 350 : 280   **b** 48 000 : 64 000   **c** 200 g : 2.5 kg   **d** 75p : £1.20 : £3

2 Write each ratio in the form 1 : n

  a 50 : 80   **b** 120 : 300   **c** 5 cm : 2 m   **d** 250 ml : 1.8 litres

3 a Share £200 in the ratio 1 : 4   **b** Share £60 in the ratio 5 : 3

  c Share £7500 in the ratio 7 : 8 : 10   **d** Share £48 in the ratio 5 : 3 : 2

4 The angles of a triangle are in the ratio 1 : 2 : 3

  Find the size of the largest angle.

5 Bronze is made by mixing copper and tin in the ratio 9 : 2

  a How much tin is mixed with 36 kg of copper?

  b How much tin is needed to make 82.5 kg of bronze?

6 Ruth, Sue and Tom serve meals at a café. Each week they share their tips in the ratio of the numbers of hours they work.

One week when Ruth works for 12 hours, Sue works for 18 hours and Tom works for 20 hours; they get £140 in tips.

How much does each person get?

7 A 3 kg tub of dishwashing powder will do 72 washes. How many washes will a 5 kg tub do?

8 Mrs Davis buys a book for each student in classes A, B and C as shown in the table.

All books cost the same amount.

Copy and complete the table.

Class	Number of students	Total Cost
A	18	£45
B	24	
C		£75

9 a In which of the tables below is y proportional to x?

Table A

x	1	2	3
y	4	3	2

Table B

x	1	2	3
y	5	10	15

Table C

x	1	2	3
y	5	6	7

  b i Use the values in the table you have chosen to draw a graph of y against x.

  ii Explain how the graph shows that y is proportional to x.

Key points

◉ A number multiplied by itself gives a **square number**, for example $3^2 = 3 \times 3 = 9$

◉ Learn these **squares**: $2^2 = 4$, $3^2 = 9$, $4^2 = 16$, $5^2 = 25$, $6^2 = 36$, $7^2 = 49$, $8^2 = 64$, $9^2 = 81$, $10^2 = 100$, $11^2 = 121$, $12^2 = 144$, $13^2 = 169$, $14^2 = 196$, $15^2 = 225$

◉ A number multiplied by itself, then multiplied by itself again gives a **cube number**, for example, $2^3 = 2 \times 2 \times 2 = 8$

◉ Learn these **cubes**: $2^3 = 8$, $3^3 = 27$, $4^3 = 64$, $5^3 = 125$, $10^3 = 1000$

◉ The **square root** of 36 is the number whose square is 36, that is $\sqrt{36} = 6$

◉ The **cube root** of 27 is the number whose cube is 27, that is $\sqrt[3]{27} = 3$

◉ An **index**, **power** or **exponent** tells you how many times the base number is multiplied by itself, for example,

5 is the index
$$2^5 = 2 \times 2 \times 2 \times 2 \times 2 = 32$$
2 is the base number

The plural of index is **indices**.

◉ On a calculator:

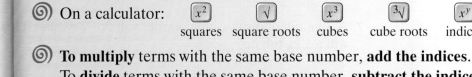

squares   square roots   cubes   cube roots   indices

◉ **To multiply** terms with the same base number, **add the indices**.
To **divide** terms with the same base number, **subtract the indices**.
To **find powers of powers**, **multiply the indices**, for example

$$5^3 \times 5^4 = 5^7 \qquad 2^5 \div 2^2 = 2^3 \qquad (7^3)^2 = 7^3 \times 7^3 = 7^6$$

Worked Example   Powers and roots

Question

Work out:

a 7^2 **b** 4^3 **c** $(-3)^2$ **d** -3^2 **e** $\sqrt{81}$ **f** $\sqrt[3]{125}$

Solution

a $7^2 = 7 \times 7 = 49$ **b** $4^3 = 4 \times 4 \times 4 = 64$ **c** $(-3)^2 = -3 \times -3 = 9$

d $-3^2 = -(3 \times 3) = -9$ **e** $\sqrt{81} = 9$ **f** $\sqrt[3]{125} = 5$

The brackets in part *c* mean multiply − 3 by itself.

In part *d* just the 3 is squared.

$\sqrt{}$ means the positive square root.

GET IT RIGHT!

Take care!
For example, 4^3 does not mean 4×3

Worked Example — Estimating roots

Question

Estimate the value of $\sqrt{61}$.

Solution

$\sqrt{49} = 7$ and $\sqrt{64} = 8$ ← *Using the square numbers above and below 61*

61 lies between 49 and 64 and is nearer to 64, so $\sqrt{61}$ is about 7.8

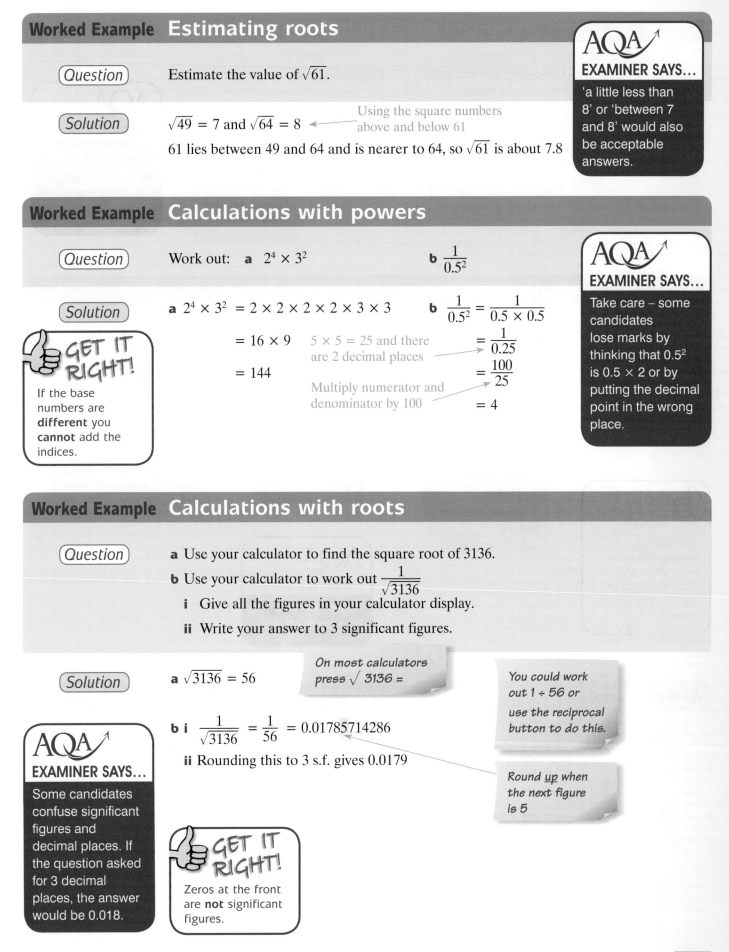

AQA EXAMINER SAYS...

'a little less than 8' or 'between 7 and 8' would also be acceptable answers.

Worked Example — Calculations with powers

Question

Work out: **a** $2^4 \times 3^2$ **b** $\dfrac{1}{0.5^2}$

Solution

a $2^4 \times 3^2 = 2 \times 2 \times 2 \times 2 \times 3 \times 3$

$= 16 \times 9$

$= 144$

b $\dfrac{1}{0.5^2} = \dfrac{1}{0.5 \times 0.5}$

$5 \times 5 = 25$ and there are 2 decimal places →

$= \dfrac{1}{0.25}$

Multiply numerator and denominator by 100 →

$= \dfrac{100}{25}$

$= 4$

GET IT RIGHT!

If the base numbers are **different** you **cannot** add the indices.

AQA EXAMINER SAYS...

Take care – some candidates lose marks by thinking that 0.5^2 is 0.5×2 or by putting the decimal point in the wrong place.

Worked Example — Calculations with roots

Question

a Use your calculator to find the square root of 3136.

b Use your calculator to work out $\dfrac{1}{\sqrt{3136}}$

  i Give all the figures in your calculator display.

  ii Write your answer to 3 significant figures.

Solution

a $\sqrt{3136} = 56$

On most calculators press $\sqrt{\ }$ 3136 =

b i $\dfrac{1}{\sqrt{3136}} = \dfrac{1}{56} = 0.01785714286$

  ii Rounding this to 3 s.f. gives 0.0179

You could work out 1 ÷ 56 or use the reciprocal button to do this.

*Round **up** when the next figure is 5*

AQA EXAMINER SAYS...

Some candidates confuse significant figures and decimal places. If the question asked for 3 decimal places, the answer would be 0.018.

GET IT RIGHT!

Zeros at the front are **not** significant figures.

Worked Example — Giving a counter example

Question

Dave says that the square of any number is always bigger than the number.

Give an example to show that Dave is wrong.

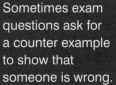

AQA EXAMINER SAYS...

Sometimes exam questions ask for a counter example to show that someone is wrong.

Solution

$0.1^2 = 0.1 \times 0.1 = 0.01$

This is smaller than 0.1, so Dave is wrong.

You could use the square of any number between 0 and 1 here.

Worked Example — Rules of indices

Question

Simplify: **a** $\dfrac{2^5 \times 2^4}{2^{12}}$ **b** $(2 \times 2^4)^3$

Remember $2 = 2^1$

Solution

a $\dfrac{2^5 \times 2^4}{2^{12}} = \dfrac{2^9}{2^{12}} = 2^{-3}$ **b** $(2 \times 2^4)^3 = (2^5)^3 = 2^{15}$

👍 **GET IT RIGHT!**

The rules of indices can be used when the base numbers or letters are the same.

To multiply, add indices.
To divide, subtract indices.

2^{-3} is equal to $\dfrac{1}{2^3} = \dfrac{1}{8}$

To find powers of powers, multiply indices, or write $(2^5)^3 = 2^5 \times 2^5 \times 2^5 = 2^{15}$

AQA EXAMINER SAYS...

Remember that $2^3 = 2 \times 2 \times 2 = 8$

Worked Example — Roots in real contexts

Question

A square rug has an area of 6.25 m².

What is the length of its sides?

Solution

Let the length of each side be x metres.

Area $= x \times x = x^2 = 6.25$

$x = \sqrt{6.25} = 2.5$

The length of each side is 2.5 metres.

Indices

END OF CHAPTER QUESTIONS

Time Yourself!

Can you complete these questions in **40** *minutes?*

1 Here is a list of numbers.

8   9   12   16   24   27   49   64   72

From this list, write down:

a the square numbers   **b** the cube numbers.

2 Write down the value of:

a the square of 6   **b** the square root of 9   **c** the square root of 100

d the cube root of 125   **e** -5^2 **f** $(-5)^2$ **g** $(-5)^3$ **h** $(-5)^4$

3 a Estimate the value of: **i** $\sqrt{15}$ **ii** $\sqrt{38}$ **iii** $\sqrt{76}$ **iv** $\sqrt{150}$ **v** $\sqrt{200}$

b Use your calculator to check your answers to part **a**.
Give values to 2 decimal places.

4 a Work out: **i** 2^6 **ii** 3^4 **iii** 7^3 **iv** 5^4 **v** 10^5

b Use your calculator to check your answers.

5 Work out: **a** $2^3 \times 3^4$ **b** $5^2 \times 2^5$ **c** $\dfrac{4^2}{2^3}$ **d** $\dfrac{7^2}{10^3}$ **e** $\dfrac{1}{0.2^2}$

6 Simplify: **a** $3 \times 3 \times 3 \times 3$ **b** $7^5 \times 7^4$ **c** $7^5 \div 7^4$ **d** $(7^5)^4$ **e** $\dfrac{5^3}{5^7}$

7 Simplify: **a** $\dfrac{6^4 \times 6}{6^2}$ **b** $(5^2 \times 5^4)^3$ **c** $\dfrac{4^7}{4^2 \times 4^3}$ **d** $\left(\dfrac{2^3}{2}\right)^5$

8 Helen is making a square poster.

She wants the area of her poster to be 6000 cm².

Find the length of the sides of the poster correct to 1 decimal place.

9 a Use your calculator to find the cube root of 2197.

b Use your calculator to work out $\dfrac{1}{\sqrt[3]{2197}}$

i Give all the figures in your calculator display.

ii Write your answer to 2 significant figures.

10 Jim is digging a hole to bury his treasure. He wants the hole to be in the shape of a cube with volume 2.5 m³.

Find the length of the sides of the hole correct to 2 decimal places.

This is a non-calculator exercise except where indicated otherwise.

1 (a) Write 56 as the product of its prime factors.

   Give your answer in index form. *(3 marks)*

   (b) Find the least common multiple (LCM) of 56 and 84. *(2 marks)*

2 p is an odd number and q is an even number.

   (a) Is $p + q$ an odd number, an even number or could it be either? *(1 mark)*

   (b) Is pq an odd number, an even number or could it be either? *(1 mark)*

3 A restaurant is open every day. It has a delivery of milk every 2 days,
   a delivery of butter every 4 days and a delivery of eggs every 7 days.
   Today they had a delivery of milk, butter and eggs.
   How many days will it be before the deliveries next **arrive** on the **same** day? *(2 marks)*

4 Use approximations to estimate the value of $\dfrac{78.32}{0.186 \times 2.19}$
   You must show all your working. *(3 marks)*

5 Calculate the value of $\dfrac{7.8 - 2.9}{12.58 - 14.49}$
   (a) Write down the full calculator display. *(1 mark)*

   (b) Give your answer to 2 significant figures. *(1 mark)*

6 Kirsty buys a bag that costs £25 to the nearest pound.

   (a) Write the least amount that she could have paid. *(1 mark)*

   (b) Write the greatest amount that she could have paid. *(1 mark)*

   AQA Spec B Foundation, Module 3, Nov 06

7 (a) Work out 400×0.8 *(1 mark)*

   (b) Work out $400 \div 0.8$ *(1 mark)*

   (c) You are told that $267 \times 35 = 9345$. Write down the value of $9345 \div 3.5$ *(1 mark)*

8 The table shows the exchange rate
   between different currencies.

   (a) Naina changes £600 into euros.

£1 (pound) is worth 1.49 euros
$1 (dollar) is worth 0.81 euros

   How many euros does she receive?

   (b) Mark changes 586 euros into dollars. How many dollars does he receive? *(2 marks)*
   (2 marks)

9 Sunita works 15 hours each week. She earns £6 per hour.

   Sunita saves $\frac{2}{5}$ of her earnings each week.

   How many weeks does it take Sunita to save £180?

   You must show all your working. *(4 marks)*

10 Karen makes matching curtains and cushions for her living room.

   For the curtains she uses $4\frac{3}{4}$ metres of material.

   For the cushions she uses $2\frac{2}{3}$ metres of material.

   How many metres of material does she use altogether? *(3 marks)*

11 Find: (a) 10% of £9.50 *(1 mark)*

   (b) 1% of £2430 *(1 mark)*

   (c) 75% of £8.40 *(2 marks)*

12 Mr and Mrs Evans are buying a television.

   How much do they save in the sale? *(2 marks)*

13 Which is the larger amount? | 80% of £30 | | $\frac{3}{5}$ of £35 |

   You **must** show your working.

   (4 marks)

14 Paul knows a quick way to work out

   15% of any amount of money.

   Use Paul's method to work out

   15% of £760

   (3 marks)

15 (a) A mobile phone is advertised as shown.

   VAT is added at 17.5%.

   Work out the cost of the mobile phone including VAT. *(3 marks)*

   (b) A shop has 32 mobile phones for sale.

   They sell 20 in one week.

   What percentage is 20 out of 32? *(2 marks)*

16 (a) Mr Little earns £18 600 per year. He gets a pay rise of 6.5%.

   After his pay rise how much does he earn?

   (3 marks)

   (b) Mrs Large earns £19 500 per year.

   After her pay rise she earns £20 670 per year.

   What was her percentage pay rise? *(3 marks)*

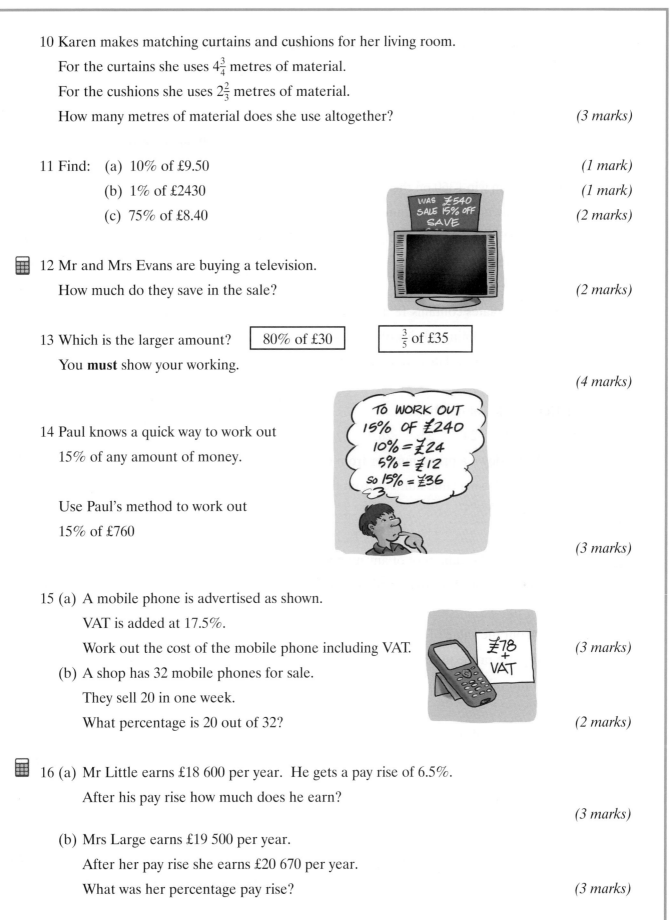

17 The ingredients needed to make 500 millilitres (ml) of a fruit drink are

orange juice 300 ml, mango juice 60 ml, lemonade 140 ml

Robert wants to make 750 ml of the fruit drink.

How much lemonade will he need? *(2 marks)*

AQA Spec B, Foundation, Module 3, Nov 06

18 Avril and Brian win £360 on the Premium Bonds.

They share the money in the ratio 1:3.

(a) How much money does each person receive? *(2 marks)*

(b) What percentage of the £360 does Brian receive? *(2 marks)*

19 A box of chocolates contains 18 milk chocolates and 12 dark chocolates.

A larger box containing 40 chocolates has the same ratio of milk to dark chocolates.

How many dark chocolates are in the larger box? *(3 marks)*

20 (a) Here is a list of numbers:

22   23   24   25   26   27   28

(i) Write down a prime number from the list. *(1 mark)*

(ii) Write down a cube number from the list. *(1 mark)*

(b) Calculate $2^5 \times 3^2$ *(2 marks)*

21 Sandy says that the cube root of any number is always smaller than the number.

Give an example to show that Sandy is wrong. *(2 marks)*

22 (a) Work out the value of $3^8 \div 3^5$ *(2 marks)*

(b) x and y are prime numbers. $xy^2 = 63$

Find the values of x and y. *(2 marks)*

23 Use your calculator to work out:

(a) 2^8 *(1 mark)*

(b) $\dfrac{1}{2.5}$ *(1 mark)*

(c) $6.4^2 + 3.18^3 + 0.95^4$

(i) Write down the full calculator display. *(1 mark)*

(ii) Write your answer to 1 significant figure. *(1 mark)*

AQA Spec B, Foundation, Module 3, Nov 06

1 p is a prime number.

Is p^2 an odd number, an even number or could it be either?
Tick the correct box.

☐ odd   ☐ even   ✓ either

AQA
EXAMINER SAYS...
You may be asked to explain your
answer to questions like this.

The only even prime number is 2.
If $p = 2$, $p^2 = 4$ is even.
All other prime numbers are odd.
When p is odd, p^2 is odd.

2 This is part of Ken's electricity bill.
How much does he pay?

Electricity Company

Present reading 9014 units   Last reading 7498 units

Each unit costs 11.6p

$9014 - 7498 = 1516$ ← The candidate is awarded one mark for subtracting the readings.

$1516 \times 11.6 = 17\ 585.6$ pence ← Another mark is awarded for multiplying by the unit cost.

$17\ 585.6 \div 100 = £175.856$ ← A third mark is awarded for dividing by 100 to change pence to £.

Answer .... £175.86 ....   (4 marks)

The final mark is awarded for giving
the answer to the nearest penny.

AQA
EXAMINER SAYS...
An answer of £175.86p would lose
the last mark.
You should not use both £ and p.

3 Sally has two cats, Kit and Kat.

Kit eats $\frac{1}{3}$ of a tin of food every day. Kat eats $\frac{1}{4}$ of a tin of food every day.

What is the **least** number of tins needed to feed the cats for eight days?

$\frac{1}{3} + \frac{1}{4} = \frac{4}{12} + \frac{3}{12} = \frac{7}{12}$ The candidate gets one mark for adding the fractions.

$8 \times \frac{7}{12} = \frac{\cancel{8}^2}{1} \times \frac{7}{\cancel{12}_3} = \frac{14}{3} = 4\frac{2}{3}$ The correct number of tins achieves the final mark.

The second mark is awarded
for multiplying by 8.

Answer .... 5 tins ....   (3 marks)

AQA
EXAMINER SAYS...
Multiplying each fraction by 8 first,
then adding the results would also
give this answer and gain full marks.

AQA
EXAMINER SAYS...
You must round **up** to the next *whole*
number of tins.

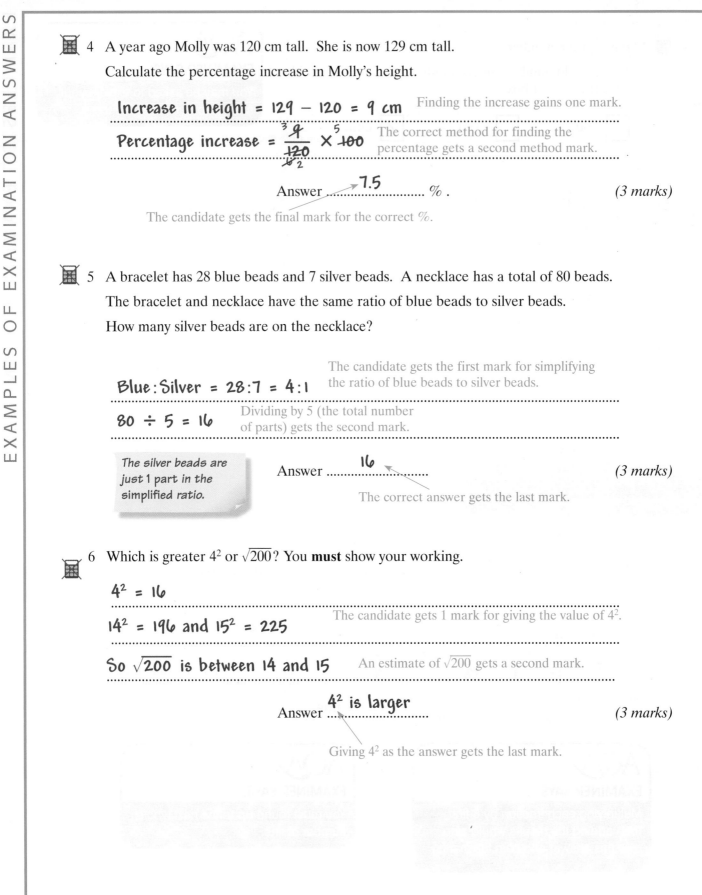

4 A year ago Molly was 120 cm tall. She is now 129 cm tall.

Calculate the percentage increase in Molly's height.

Increase in height = 129 − 120 = 9 cm   Finding the increase gains one mark.

Percentage increase = $\frac{9}{120} \times 100$ The correct method for finding the percentage gets a second method mark.

Answer ....7.5............ % .   *(3 marks)*

The candidate gets the final mark for the correct %.

5 A bracelet has 28 blue beads and 7 silver beads. A necklace has a total of 80 beads.

The bracelet and necklace have the same ratio of blue beads to silver beads.

How many silver beads are on the necklace?

The candidate gets the first mark for simplifying the ratio of blue beads to silver beads.

Blue : Silver = 28 : 7 = 4 : 1

80 ÷ 5 = 16   Dividing by 5 (the total number of parts) gets the second mark.

The silver beads are just 1 part in the simplified ratio.

Answer ....16............   *(3 marks)*

The correct answer gets the last mark.

6 Which is greater 4^2 or $\sqrt{200}$? You **must** show your working.

$4^2 = 16$

$14^2 = 196$ and $15^2 = 225$ The candidate gets 1 mark for giving the value of 4^2.

So $\sqrt{200}$ is between 14 and 15   An estimate of $\sqrt{200}$ gets a second mark.

Answer ....4^2 is larger....   *(3 marks)*

Giving 4^2 as the answer gets the last mark.

1 Angles

Key words

- acute angle
- obtuse angle
- reflex angle
- opposite angles
- alternate angles
- corresponding angles
- bearing
- isosceles triangle or trapezium
- equilateral triangle
- right-angled triangle
- quadrilateral
- square
- rectangle
- kite
- trapezium (pl. trapezia)
- parallelogram
- rhombus
- regular polygon
- pentagon
- hexagon
- octagon
- interior angle
- exterior angle

Key points

Angles

- **Angles on a straight line** add up to 180°.

- **Angles at a point** add up to 360°.

- **Opposite angles** are equal.

- **Corresponding angles** are equal.

- **Alternate angles** are equal.

Bearings

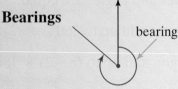

bearing

Measure the angle clockwise from north and write it as a three figure number, for example, 300°, 120°, 045°.

Triangles

- **Isosceles triangle** – two equal sides, two equal base angles.

- **Equilateral triangle** – three equal sides, three equal angles. Each angle is 60°

- **Right-angled** triangle – one angle equals 90°.

- **Angles in a triangle** add up to 180°.

- **Exterior angle** of a triangle equals the sum of the interior opposite angles.

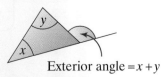

Exterior angle = $x + y$

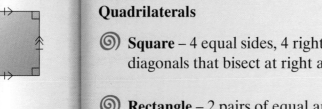

Quadrilaterals

⊚ **Square** – 4 equal sides, 4 right angles, 2 pairs of parallel sides, equal diagonals that bisect at right angles.

⊚ **Rectangle** – 2 pairs of equal and parallel sides, 4 right angles, equal diagonals that bisect each other.

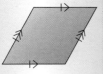

⊚ **Rhombus** – 4 equal sides, 2 pairs of parallel sides, diagonals bisect at right angles.

⊚ **Parallelogram** – 2 pairs of equal and parallel sides, diagonals bisect each other.

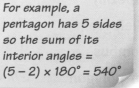

⊚ **Kite** – 2 pairs of equal adjacent sides, long diagonal bisects shorter one at right angles.

⊚ **Isosceles trapezium** – 1 pair of parallel sides, other pair equal, diagonals equal.

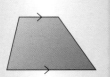

⊚ **Trapezium** – 1 pair of parallel sides.

> *You need to be able to describe a quadrilateral or name a quadrilateral from its properties.*

Polygons

⊚ **Regular polygon** – all sides and angles are equal.

⊚ **Exterior angles** of a polygon add up to 360°.

⊚ **Interior angles** of a polygon add up to $(n - 2) \times 180°$, where n is the number of sides.

> *For example, a regular octagon has 8 equal sides and angles so each exterior angle = $\frac{360°}{8}$ = 45°*

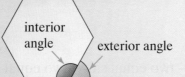

interior angle

exterior angle

> *For example, a pentagon has 5 sides so the sum of its interior angles = $(5 - 2) \times 180° = 540°$*

exterior angle + interior angle = 180°

THE GRADE To get a grade **C** you need to be able to classify a quadrilateral by its geometric properties and calculate exterior and interior angles of a regular polygon.

Worked Example   Bearings

(Question)   The diagram shows a map of two routes from airport A.

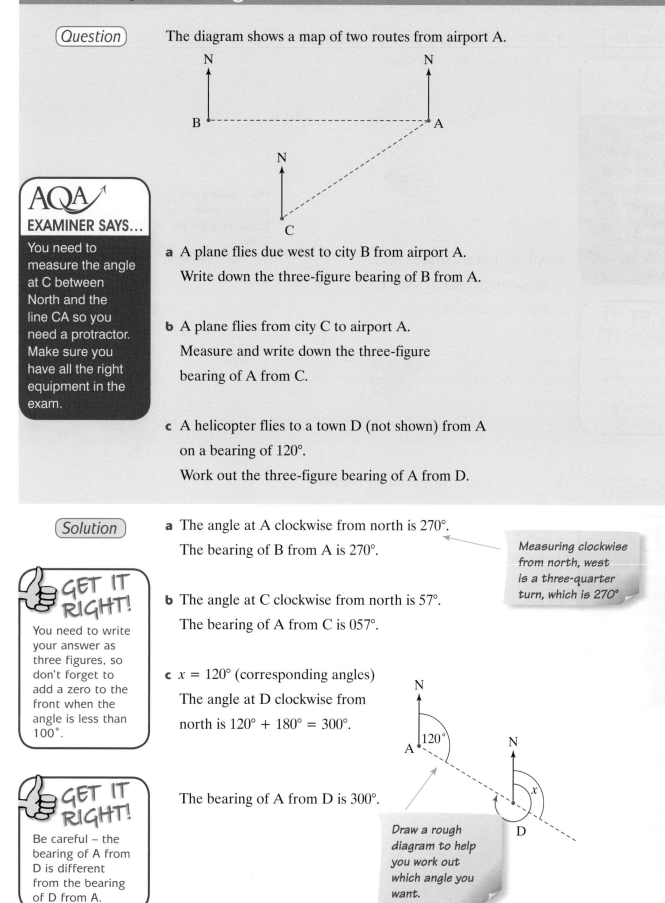

AQA

EXAMINER SAYS...

You need to measure the angle at C between North and the line CA so you need a protractor. Make sure you have all the right equipment in the exam.

a   A plane flies due west to city B from airport A.

Write down the three-figure bearing of B from A.

b   A plane flies from city C to airport A.

Measure and write down the three-figure

bearing of A from C.

c   A helicopter flies to a town D (not shown) from A

on a bearing of 120°.

Work out the three-figure bearing of A from D.

(Solution)   a   The angle at A clockwise from north is 270°.

The bearing of B from A is 270°.

Measuring clockwise from north, west is a three-quarter turn, which is 270°

GET IT RIGHT!

You need to write your answer as three figures, so don't forget to add a zero to the front when the angle is less than 100°.

b   The angle at C clockwise from north is 57°.

The bearing of A from C is 057°.

c $x = 120°$ (corresponding angles)

The angle at D clockwise from

north is 120° + 180° = 300°.

The bearing of A from D is 300°.

Draw a rough diagram to help you work out which angle you want.

GET IT RIGHT!

Be careful – the bearing of A from D is different from the bearing of D from A.

Worked Example Angles

a Work out the size of angles a, b and c.

Give reasons for your answers.

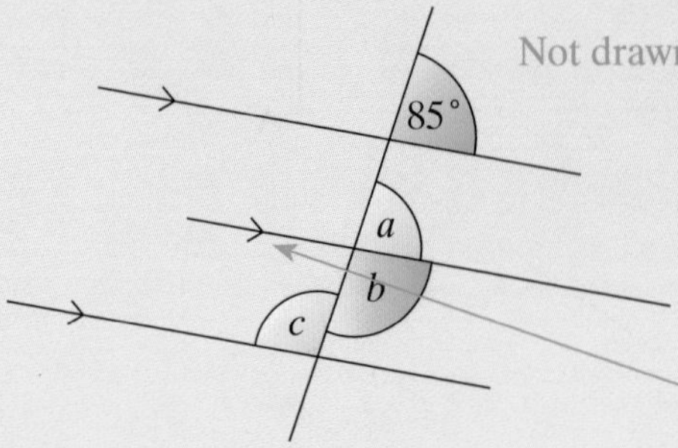

Not drawn accurately

85°

a

b

c

The arrows show that the three lines in the diagram are parallel.

b AB and CD are straight lines.

Work out the size of angle x.

Show all your working.

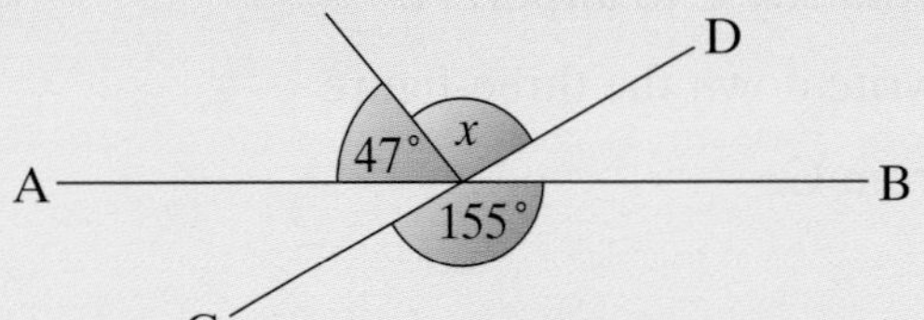

D

47° *x*

A — — B

155°

C   Not drawn accurately

a $a = 85°$ (corresponding angles)

$a + b = 180°$ (angles on a straight line)
$85° + b = 180°$
$b = 180° - 85° = 95°$

It's easy to miss alternate angles when they are obtuse or when the parallel lines are not horizontal.

$c = b = 95°$ (alternate angles)

b $x + 47° = 155°$ (opposite angles)

$x = 155° - 47° = 108°$

x is not an opposite angle to 155°
$x + 47°$ is the angle made by AB and CD crossing.

Angles ▦

END OF CHAPTER QUESTIONS

Time Yourself!

Can you complete these questions in **25** minutes?

1 Look at this list of words.

| Acute | Right | Full-turn | Obtuse | Half-turn | Reflex |

Choose the correct word to describe each of the following angles.

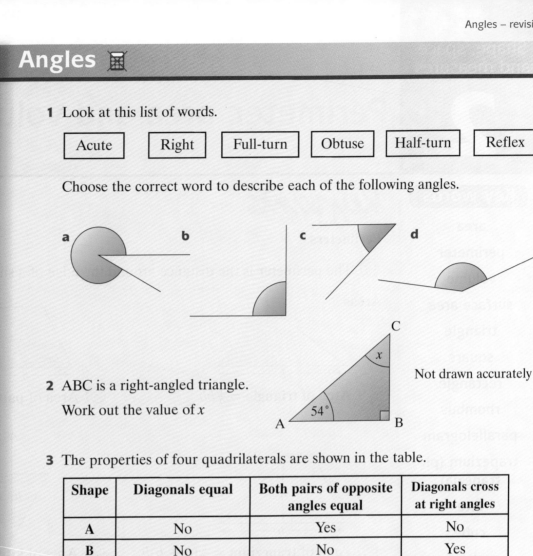

2 ABC is a right-angled triangle.

Work out the value of x

Not drawn accurately

3 The properties of four quadrilaterals are shown in the table.

Shape	Diagonals equal	Both pairs of opposite angles equal	Diagonals cross at right angles
A	No	Yes	No
B	No	No	Yes
C	Yes	Yes	Yes
D			

a A square is described by shape ....

b A kite is described by shape ....

c Shape **D** is a rectangle. Complete the table.

4 The lines AB and CD are parallel.

a Find the value of x.

b Find the value of y.

c Find the value of z.

Give a reason for each of your answers.

Not drawn accurately

5 ABCDE is a regular pentagon.

a Work out the value of x.

b Calculate the exterior angle of the regular pentagon, y.

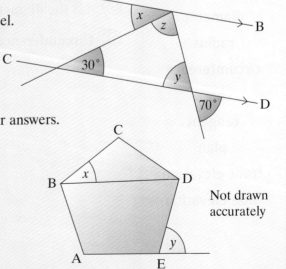

Not drawn accurately

Perimeter, area and volume

Key points

Perimeters

⊚ The **perimeter** is the distance around the edge of a shape.

Areas

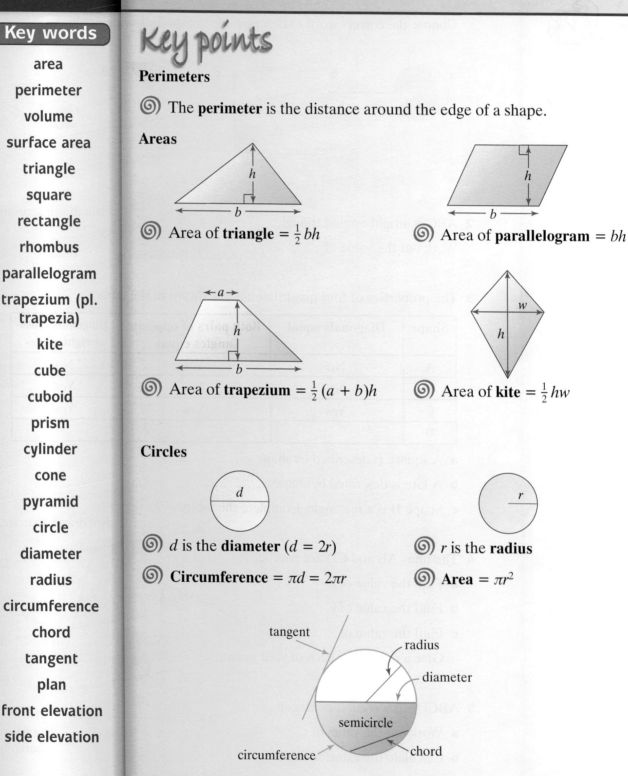

⊚ Area of **triangle** $= \frac{1}{2}bh$

⊚ Area of **parallelogram** $= bh$

⊚ Area of **trapezium** $= \frac{1}{2}(a + b)h$

⊚ Area of **kite** $= \frac{1}{2}hw$

Circles

⊚ d is the **diameter** $(d = 2r)$

⊚ r is the **radius**

⊚ **Circumference** $= \pi d = 2\pi r$

⊚ **Area** $= \pi r^2$

3-D solids

Shape	Volume	Surface area

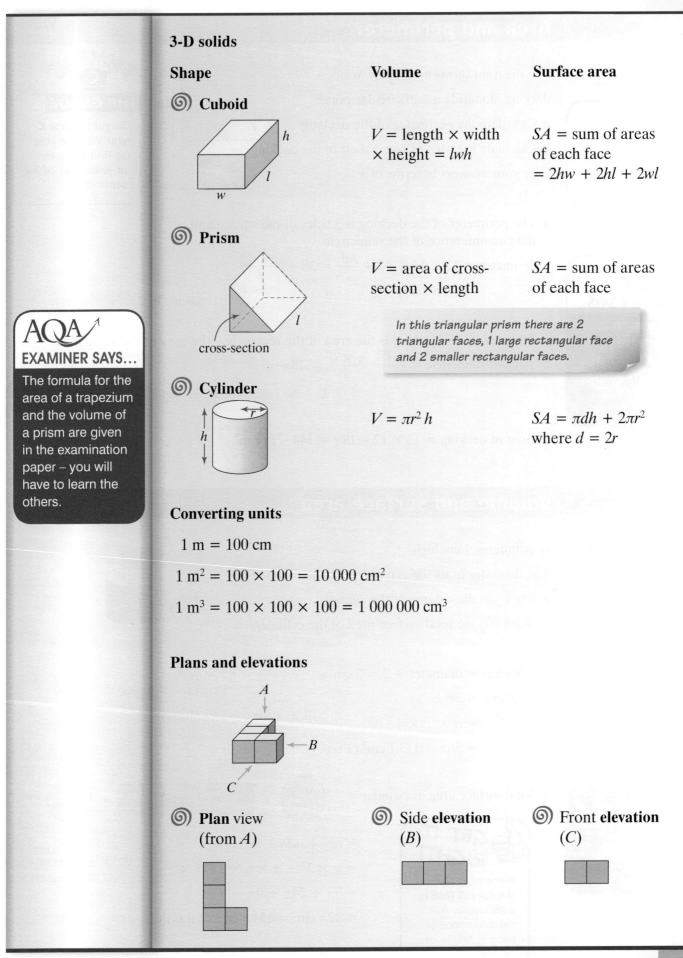

Cuboid

V = length × width × height = lwh

SA = sum of areas of each face = $2hw + 2hl + 2wl$

Prism

V = area of cross-section × length

SA = sum of areas of each face

cross-section

> In this triangular prism there are 2 triangular faces, 1 large rectangular face and 2 smaller rectangular faces.

Cylinder

$V = \pi r^2 h$

$SA = \pi dh + 2\pi r^2$ where $d = 2r$

Converting units

1 m = 100 cm

1 m^2 = 100 × 100 = 10 000 cm^2

1 m^3 = 100 × 100 × 100 = 1 000 000 cm^3

Plans and elevations

Plan view (from A)

Side **elevation** (B)

Front **elevation** (C)

Worked Example Area and perimeter

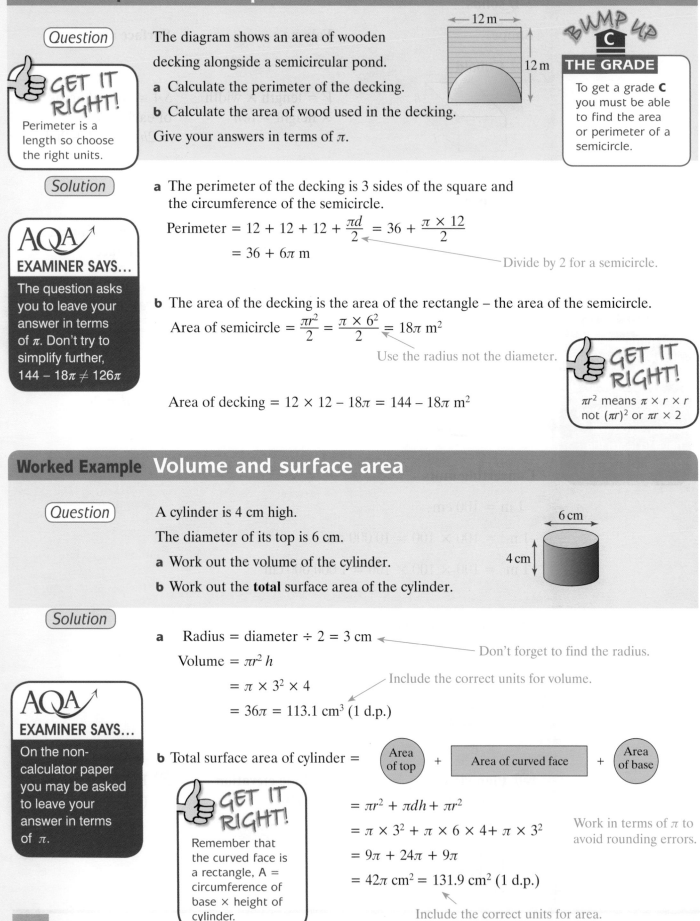

Question

GET IT RIGHT!

Perimeter is a length so choose the right units.

The diagram shows an area of wooden decking alongside a semicircular pond.

a Calculate the perimeter of the decking.

b Calculate the area of wood used in the decking.

Give your answers in terms of π.

←——12 m——→

12 m

BUMP UP

C

THE GRADE

To get a grade **C** you must be able to find the area or perimeter of a semicircle.

Solution

AQA✓

EXAMINER SAYS...

The question asks you to leave your answer in terms of π. Don't try to simplify further, $144 - 18\pi \neq 126\pi$

a The perimeter of the decking is 3 sides of the square and the circumference of the semicircle.

$$\text{Perimeter} = 12 + 12 + 12 + \frac{\pi d}{2} = 36 + \frac{\pi \times 12}{2}$$

$$= 36 + 6\pi \text{ m}$$

Divide by 2 for a semicircle.

b The area of the decking is the area of the rectangle – the area of the semicircle.

$$\text{Area of semicircle} = \frac{\pi r^2}{2} = \frac{\pi \times 6^2}{2} = 18\pi \text{ m}^2$$

Use the radius not the diameter.

GET IT RIGHT!

πr^2 means $\pi \times r \times r$ not $(\pi r)^2$ or $\pi r \times 2$

$$\text{Area of decking} = 12 \times 12 - 18\pi = 144 - 18\pi \text{ m}^2$$

Worked Example Volume and surface area

Question

A cylinder is 4 cm high.

The diameter of its top is 6 cm.

a Work out the volume of the cylinder.

b Work out the **total** surface area of the cylinder.

6 cm

4 cm

Solution

AQA✓

EXAMINER SAYS...

On the non-calculator paper you may be asked to leave your answer in terms of π.

a Radius = diameter ÷ 2 = 3 cm

Don't forget to find the radius.

Volume = $\pi r^2 h$

Include the correct units for volume.

$$= \pi \times 3^2 \times 4$$

$$= 36\pi = 113.1 \text{ cm}^3 \text{ (1 d.p.)}$$

GET IT RIGHT!

Remember that the curved face is a rectangle, A = circumference of base × height of cylinder.

b Total surface area of cylinder = ⬭ Area of top + ▭ Area of curved face + ⬭ Area of base

$$= \pi r^2 + \pi d h + \pi r^2$$

$$= \pi \times 3^2 + \pi \times 6 \times 4 + \pi \times 3^2$$

$$= 9\pi + 24\pi + 9\pi$$

$$= 42\pi \text{ cm}^2 = 131.9 \text{ cm}^2 \text{ (1 d.p.)}$$

Work in terms of π to avoid rounding errors.

Include the correct units for area.

Worked Example Plans and elevations

(Question)

This 3-D solid is made from six cubes.

If you find this difficult, try making the shape and turning it around so that you can see it from different sides. You could try making some other shapes to practise.

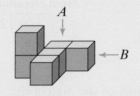

GET IT RIGHT!

A plan is a bird's eye view, so imagine you are looking directly down from above.

a On squared paper draw the plan from A.

b On squared paper draw the elevation from B.

(Solution)

GET IT RIGHT!

An elevation is a view from one of the sides. Make sure you include all the squares that you can see.

a Plan

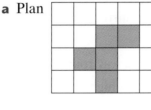

b Elevation

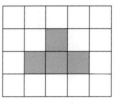

BUMP UP

C

THE GRADE To get a grade **C** you need to be able to find the volume and surface area of prisms and cylinders (as in the next example).

Perimeter, area and volume

END OF CHAPTER QUESTIONS

Time Yourself!

Can you complete these questions in **40** minutes?

1 Write down the names of these 3-D solids.

a **b** **c** **d**

2 a This T-shape is made of rectangles.

10.4 cm

2.2 cm 2.2 cm

4.1 cm

8.5 cm

2.2 cm

Not drawn accurately

Calculate the area of the T-shape.

b This shape is made of two semicircles.

The radius of the larger semicircle is 6 cm.

Calculate the perimeter of the shape.

6 cm

3 A matchbox is in the shape of a cuboid.

The diagrams show the plan view, front elevation and side elevation.

Plan view

5 cm

7 cm

Front elevation

2 cm

7 cm

Side elevation

2 cm

5 cm

Calculate the volume of the matchbox.

4 A semicircular patio has a radius of 2.5 m.

Calculate the area of the patio. State the units of your answer.

5 The diagram shows a rectangular mirror
with a wooden frame around it.

The frame is the same width all the way round.

The mirror is 20 cm wide and 25 cm high.

The total width of the mirror **and** frame is 30 cm.

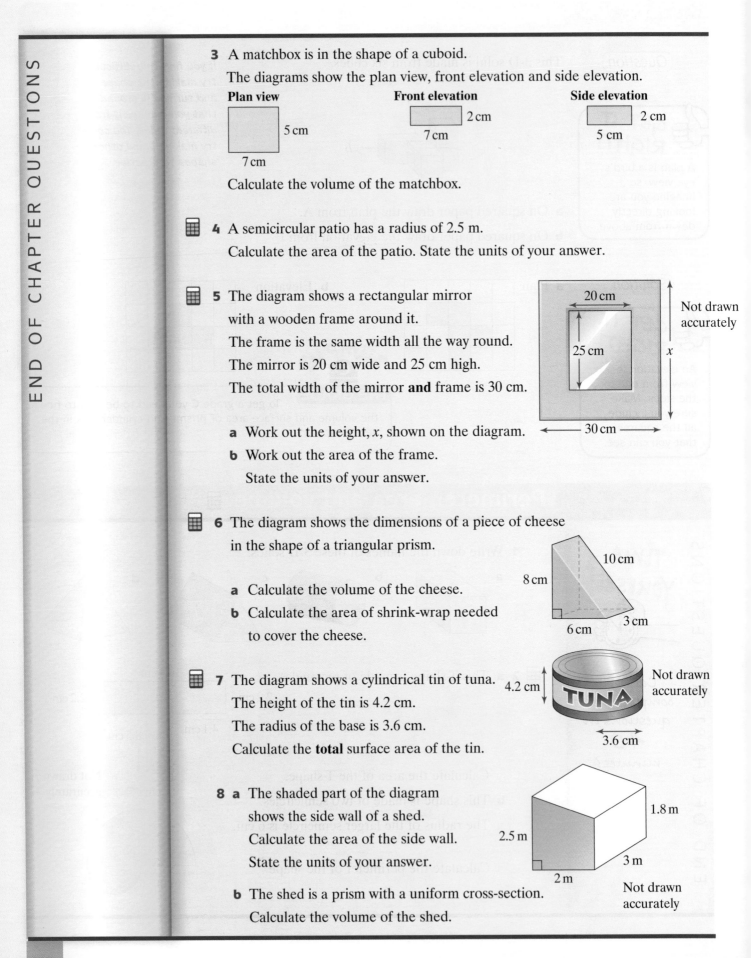

20 cm

25 cm

x

30 cm

Not drawn
accurately

a Work out the height, x, shown on the diagram.

b Work out the area of the frame.
State the units of your answer.

6 The diagram shows the dimensions of a piece of cheese
in the shape of a triangular prism.

a Calculate the volume of the cheese.

b Calculate the area of shrink-wrap needed
to cover the cheese.

10 cm

8 cm

6 cm

3 cm

7 The diagram shows a cylindrical tin of tuna.

The height of the tin is 4.2 cm.

The radius of the base is 3.6 cm.

Calculate the **total** surface area of the tin.

4.2 cm

TUNA

Not drawn
accurately

3.6 cm

8 a The shaded part of the diagram
shows the side wall of a shed.
Calculate the area of the side wall.
State the units of your answer.

1.8 m

2.5 m

2 m

3 m

Not drawn
accurately

b The shed is a prism with a uniform cross-section.
Calculate the volume of the shed.

Reflections and rotations

Key words

symmetry
(reflection)
(rotation)

line of symmetry

order of rotation
symmetry

image

transformation

mapping

reflection

rotation

centre of rotation

Key points

🌀 A **transformation** changes the **position** or **size** of a shape.

🌀 A shape has **reflection symmetry** in **a line of symmetry** if reflecting it in that line gives an identical shape.

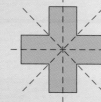 

This shape doesn't have any lines of symmetry: check by drawing and folding.

4 lines of symmetry   1 line of symmetry   0 lines of symmetry

🌀 A shape has **rotation symmetry** if a rotation about its centre through an angle greater than 0° and less than 360° gives an identical-looking shape. The **order of rotation symmetry** is the number of ways a shape would fit on top of itself as it is rotated through 360°.

rotation symmetry

order 4   order 1   order 2

🌀 A **reflection** is a transformation involving a **mirror line** (or **axis of symmetry**), in which the line from each point to its image is perpendicular to the mirror line and has its midpoint on the mirror line.

To describe a reflection fully you must describe the position, or give the equation, of its mirror line.

Triangle A has been reflected in the mirror line $y = 1$ to give the image B.

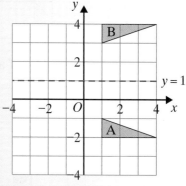

A **rotation** is a transformation in which the shape is turned about a fixed point, the **centre of rotation**. To describe a rotation fully, you must give the centre, angle and direction.

A positive angle is anticlockwise and a negative angle is clockwise.

Triangle A is rotated about the origin through 90° anticlockwise to give the image C.

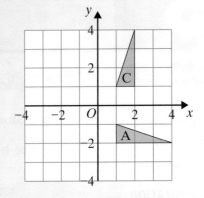

Worked Example Symmetry

(Question)

a Shade **two** more squares to make a pattern with rotation symmetry of order 2 and centre ✗.

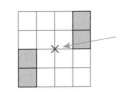

b Shade **three** more squares to make the grid have a pattern with 1 line of symmetry.

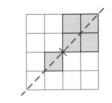

(Solution)

a

Use tracing paper to help. Put your pencil point on the ✗ while you turn the paper.

GET IT RIGHT!

You are looking for rotation symmetry in this part so don't think about lines of symmetry.

b

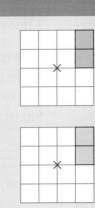

There are lots of alternative solutions to **b**.

For example,

AQA EXAMINER SAYS...

The question says that the whole grid must have 1 line of symmetry – not just the pattern of squares. You must use exactly 3 squares.

THE GRADE
To get a grade **C** you must be able to reflect shapes in the lines $y = x$ and $y = -x$. You will also need to rotate shapes about any point and combine transformations. Work through the following example and make sure you know how to **fully** describe reflections and rotations.

Worked Example   Reflections and rotations

(Question)

AQA
EXAMINER SAYS...
When a question tells you to label your images, make sure that you do otherwise the examiner might not be able to tell which image is which.

(Solution)

GET IT RIGHT!
Remember that $x = 1$ is a vertical line.

The diagram shows a shaded triangle.

a Reflect the triangle in the line $x = 1$
Label the image A.

b Reflect your image, triangle A, in the line $y = x$.
Label the image B.

c Describe fully the **single** transformation that maps the original shaded triangle onto triangle B.

a

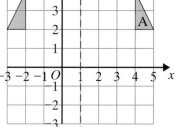

b

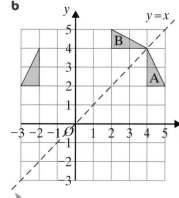

GET IT RIGHT!
Be extra careful when reflecting in $y = x$. Turn your page so that the mirror line is vertical as this will help you draw the image.

c

The transformation is a rotation 90° clockwise about (1, 1).

Draw in the mirror line first and use tracing paper to help.

Use tracing paper to help. If you are not sure about the centre of rotation you will still score marks for describing the angle and direction.

AQA
EXAMINER SAYS...
The question tells you to reflect triangle A – don't use the wrong triangle!

AQA
EXAMINER SAYS...
To describe a rotation you must give the centre, angle and direction.

Reflections and rotations

Time Yourself!

Can you complete these questions in **35** minutes?

1 a This quadrilateral has **exactly** two lines of symmetry.

   i Copy the diagram and draw in the lines of symmetry.

   ii What is the order of rotation symmetry of this quadrilateral?

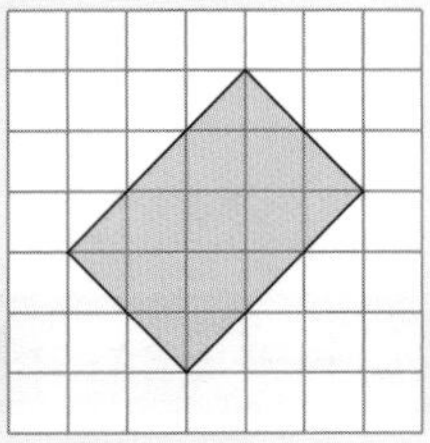

b Sarah is drawing a quadrilateral with **exactly** one line of symmetry.

   i Copy and complete the diagram to show a quadrilateral that Sarah could draw.

   ii Write down the name of Sarah's quadrilateral.

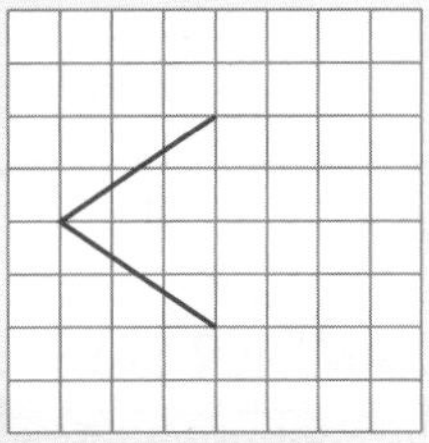

2 Copy the shape three times.

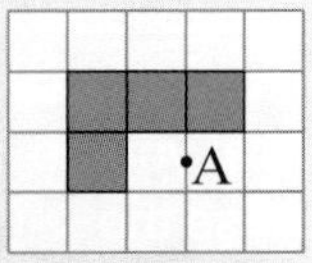

   a On your first diagram shade one more square so that the shaded shape has **exactly** one line of symmetry.

   b On your second diagram shade one square so that the shaded shape has rotation symmetry of order 2.

   c On your third diagram show the position of the shape after a rotation of 180° about the point A.

3 The diagram shows two triangles.

   a Describe fully the single transformation that maps triangle A onto triangle B.

   b Copy the diagram and reflect triangle B in the line $y = -1$.

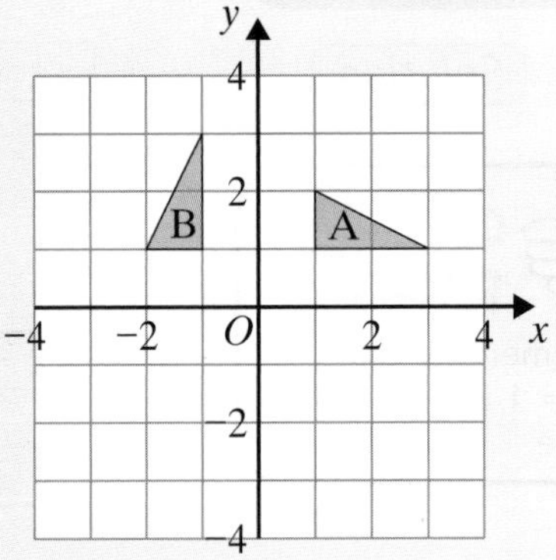

4 The length, width and height of a cuboid are all different.
A plane cuts each cuboid into two equal parts.
Which of these are planes of symmetry?

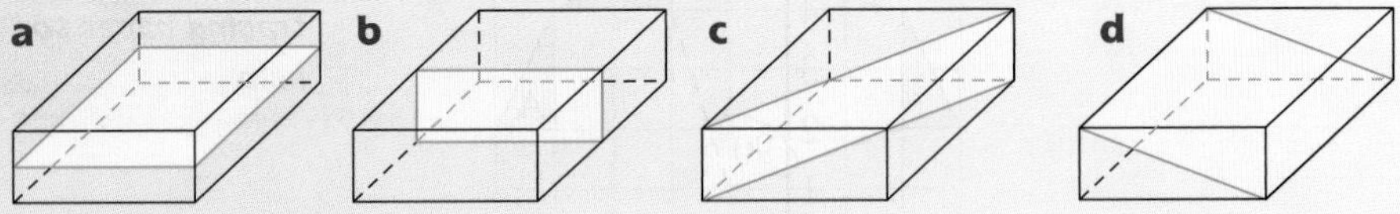

   a   **b**   **c**   **d**

5 The diagram shows a shaded flag.
Copy the diagram.

   a Rotate the shaded flag 90° clockwise about the origin.
   Label this new flag with the letter A.

   b Reflect the original shaded flag in the line $x = -1$.
   Label this new flag with the letter B.

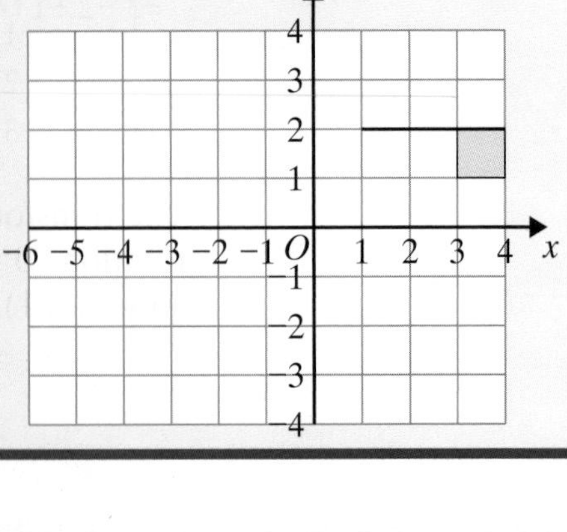

Translations and enlargements

Key words

object

image

mapping

congruent

similar

scale factor

transformation

enlargement

centre of enlargement

translation

vector

A scale factor of 1 doesn't change the shape.
A scale factor greater than 1 enlarges the shape.
A scale factor between 0 and 1 shrinks the shape.

Key points

○ A **translation** is a transformation in which every point of the **object** moves the same distance and in the same direction to map onto the **image**.

Triangle A has been mapped onto triangle B by a translation of 3 units to the right and 2 units down.

The object and the image are congruent because they are the same size and shape.

The **translation vector** to map A to B is $\begin{pmatrix} 3 \\ -2 \end{pmatrix}$

To return triangle B to triangle A use a translation of 3 units to the left and 2 units up.

The **translation vector** to map B to A is $\begin{pmatrix} -3 \\ 2 \end{pmatrix}$

The top number of a translation vector is the movement across (positive to the right, negative to the left).

The bottom number of a translation vector is the vertical movement (positive upwards, negative downwards).

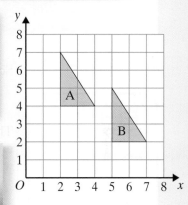

○ An **enlargement** is a transformation that changes the size of an object but not its shape. To describe an enlargement fully you must give the **centre of enlargement** and the **scale factor**.

Triangle B is an enlargement of triangle A.
The **centre of enlargement** is (1, 7).
The **scale factor** is 3.

Scale factor = length of line on enlargement / length of line on original

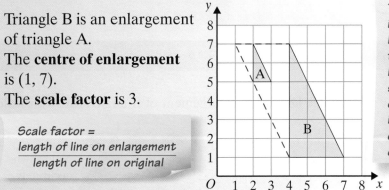

The object and the image are similar because they are the same shape but a different size. Each side of triangle B is 3 times the corresponding side of triangle A.

○ **Congruent** shapes are the same shape and size.

○ **Similar** shapes are the same shape but different sizes. Their corresponding angles are equal and their corresponding sides are in the same ratio.

○ Maps and scales – a map scale of 1 : 250 000 means that 1 cm on the map represents 250 000 cm in real life.
250 000 cm = 2 500 m = 2.5 km so 1 cm represents 2.5 km.

The units must match here.

THE GRADE   To get a grade **C** you need to understand enlargements with fractional scale factors.

Worked Example   Enlargements

(Question)   The diagram shows two triangles.

Describe fully the single transformation that maps triangle A onto triangle B.

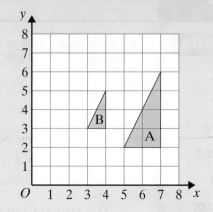

(Solution)

Scale factor = $\dfrac{\text{length on enlargement}}{\text{length on original}}$

$= \dfrac{1}{2}$

The transformation is an enlargement by scale factor $\frac{1}{2}$ with centre of enlargement (1, 4).

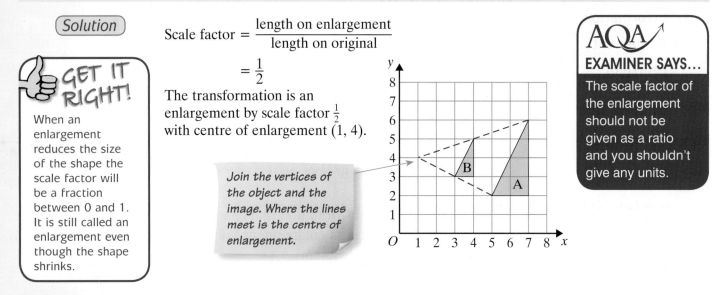

Join the vertices of the object and the image. Where the lines meet is the centre of enlargement.

GET IT RIGHT!

When an enlargement reduces the size of the shape the scale factor will be a fraction between 0 and 1. It is still called an enlargement even though the shape shrinks.

AQA
EXAMINER SAYS...
The scale factor of the enlargement should not be given as a ratio and you shouldn't give any units.

Worked Example   Similar shapes

(Question)   Triangle DEF is an enlargement of triangle ABC with scale factor $\frac{3}{2}$.

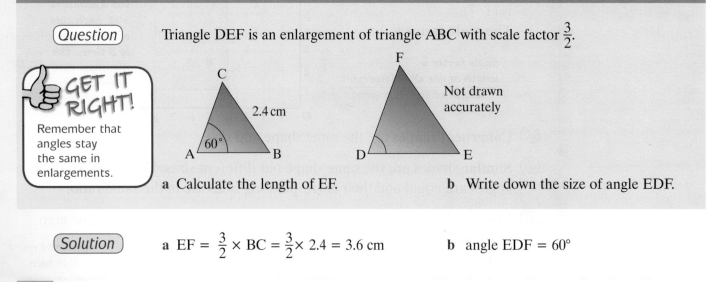

Not drawn accurately

GET IT RIGHT!

Remember that angles stay the same in enlargements.

a Calculate the length of EF.   **b** Write down the size of angle EDF.

(Solution)   **a** EF $= \frac{3}{2} \times$ BC $= \frac{3}{2} \times 2.4 = 3.6$ cm   **b** angle EDF $= 60°$

THE GRADE

To get a grade **C** you need to be able to use translation vectors. You will also need to combine transformations. Practise the following type of question to get confident.

Worked Example Combining transformations

(Question)

The diagram shows a shaded triangle.

a Reflect the triangle in the line $y = x$.
Label the image A.

b Rotate triangle **A** 90° clockwise about the point (–1, 2).
Label the image B.

c Reflect triangle **B** in the line $y = 0$
Label the image C.

d Describe fully the **single** transformation that maps the original shaded triangle onto triangle C.

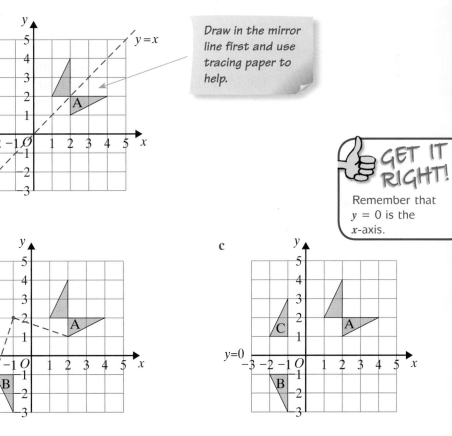

EXAMINER SAYS...

When a question tells you to label your images, make sure that you do otherwise the examiner might not be able to tell which image is which.

(Solution)

GET IT RIGHT!

Be extra careful when reflecting in $y = x$. Turn your page so that the mirror line is vertical as this will help you draw the image.

EXAMINER SAYS...

Make sure you use the given centre of rotation and not the origin.

Use tracing paper to help. Put your pencil point on (–1, 2) and turn the paper through a quarter turn.

Draw in the mirror line first and use tracing paper to help.

GET IT RIGHT!

Remember that $y = 0$ is the x-axis.

d The original shape is translated 3 units left and 1 unit down, so the transformation is a translation of $\begin{pmatrix} -3 \\ -1 \end{pmatrix}$.

Remember the top number is the movement across – it is negative here to show it is to the left.

Remember the bottom number is the vertical movement – it is negative here to show it is downwards.

Translations and enlargements

Time Yourself!

Can you complete these questions in **30** minutes?

1 The diagram shows a shaded shape on a centimetre grid.

   a Copy the diagram and translate the shaded shape 2 units across and 1 unit down.

   b Enlarge the shaded shape, from centre of enlargement, A, by a scale factor of 3.

   c How many times bigger is the area of the enlarged shape than the area of the original shape?

2 Square B is an enlargement of square A. The squares are drawn on a centimetre grid.

   a What is the scale factor of the enlargement?

   b Square **B** is enlarged by scale factor 5 to give square C. Write down the length of a side of square C.

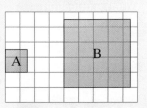

3 The scale of a model boat is 1 : 40.

   a The width of the model boat is 240 mm. Calculate the actual width of the boat in metres.

   b The actual length of the boat is 44 m. Calculate the length of the model boat in centimetres.

4 Isosceles trapezium Y is an enlargement of isosceles trapezium X.

   a Work out the ratio of corresponding lengths.

   b Calculate the side, a, of trapezium Y.

   c Calculate the sloping height, b, of trapezium X.

   d Write down the ratio of the perimeters of the trapezia in its simplest form.

Not drawn accurately

5 Copy the axes and shaded shape X.

   a Translate the shaded shape X by the vector $\begin{pmatrix} 3 \\ -1 \end{pmatrix}$.

   b Write down the translation vector that would return the image of X back to its original position.

   Now copy the axes and shaded shape Y.

   c Draw the enlargement of Y with scale factor $\frac{1}{2}$ and centre of enlargement $(0, 0)$.

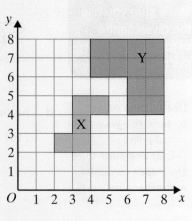

5 Measures and drawing

Key words

speed

density

lower bound

upper bound

perpendicular
bisector

angle bisector

net

locus (pl. loci)

equidistant

Key points

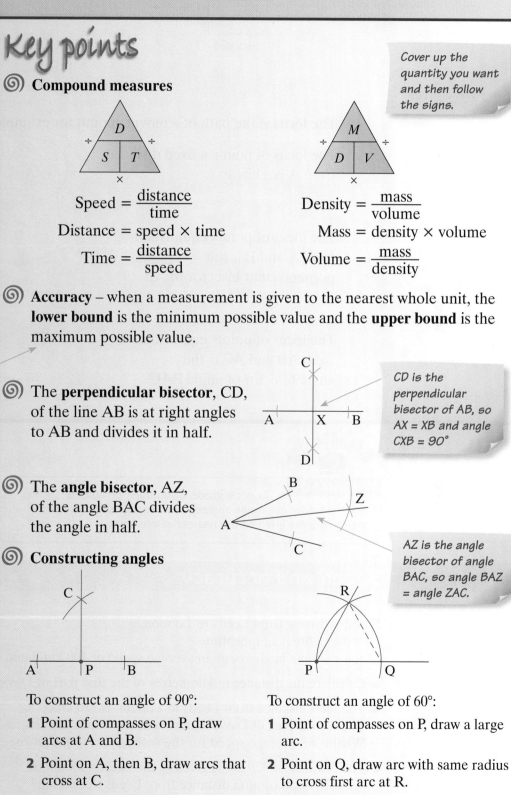

🌀 **Compound measures**

$$\text{Speed} = \frac{\text{distance}}{\text{time}}$$

$$\text{Distance} = \text{speed} \times \text{time}$$

$$\text{Time} = \frac{\text{distance}}{\text{speed}}$$

$$\text{Density} = \frac{\text{mass}}{\text{volume}}$$

$$\text{Mass} = \text{density} \times \text{volume}$$

$$\text{Volume} = \frac{\text{mass}}{\text{density}}$$

Cover up the quantity you want and then follow the signs.

🌀 **Accuracy** – when a measurement is given to the nearest whole unit, the **lower bound** is the minimum possible value and the **upper bound** is the maximum possible value.

If a length is 5 cm correct to the nearest centimetre, the lower bound = 4.5 cm and the upper bound = 5.5 cm

🌀 The **perpendicular bisector**, CD, of the line AB is at right angles to AB and divides it in half.

CD is the perpendicular bisector of AB, so AX = XB and angle CXB = 90°

🌀 The **angle bisector**, AZ, of the angle BAC divides the angle in half.

AZ is the angle bisector of angle BAC, so angle BAZ = angle ZAC.

🌀 **Constructing angles**

Revise how to construct perpendicular and angle bisectors using a ruler and compasses.

To construct an angle of 90°:

1 Point of compasses on P, draw arcs at A and B.

2 Point on A, then B, draw arcs that cross at C.

3 Join CP.

To construct an angle of 60°:

1 Point of compasses on P, draw a large arc.

2 Point on Q, draw arc with same radius to cross first arc at R.

3 Join PR.

A **net** is a two-dimensional shape made of polygons that can be folded to make a three-dimensional solid, for example,

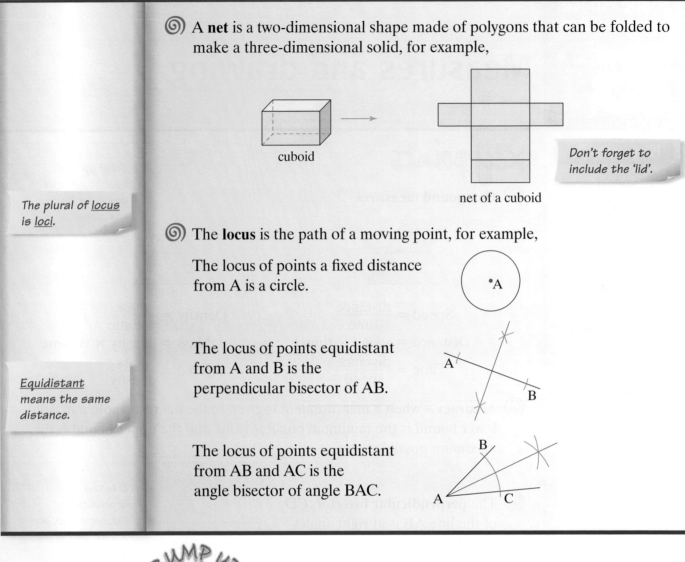

cuboid

net of a cuboid

Don't forget to include the 'lid'.

The **locus** is the path of a moving point, for example,

The plural of <u>locus</u> is <u>loci</u>.

The locus of points a fixed distance from A is a circle.

The locus of points equidistant from A and B is the perpendicular bisector of AB.

<u>Equidistant</u> means the same distance.

The locus of points equidistant from AB and AC is the angle bisector of angle BAC.

THE GRADE   To get a grade **C** you must be able to answer harder questions on speed and recognise accuracy in measurements given to the nearest whole unit.

Worked Example   Speed and accuracy

(Question)

David is driving from Leeds to London.
He takes a break at lunchtime.
In the morning he drives at an average speed of 100 km/h and takes 1 hour 45 minutes.

a Calculate the distance in kilometres of the first part of David's journey.

b The **total** distance from Leeds to London is 320 km.
The second part of David's journey takes 1 hour 15 minutes.
What is his average speed for the second part of his journey?

c The distance of 320 km is given to the nearest kilometre.
What is the maximum distance from Leeds to London?

Solution

a Time = 1 hour 45 min = $1\frac{3}{4}$ h = 1.75 h

distance = speed × time

distance = 100 × 1.75 = 175 km

b Distance of second part = 320 − 175 = 145 km

Time of second part = 1 hour 15 min = $1\frac{1}{4}$ h = 1.25 h

speed = $\dfrac{\text{distance}}{\text{time}}$

speed = $\dfrac{145}{1.25}$ = 116 km/h

c The maximum distance from Leeds to London is 320.5 km

Worked Example Constructions

Question

Construct the angle bisector of angle BAC.
You **must** show your construction arcs.

Solution

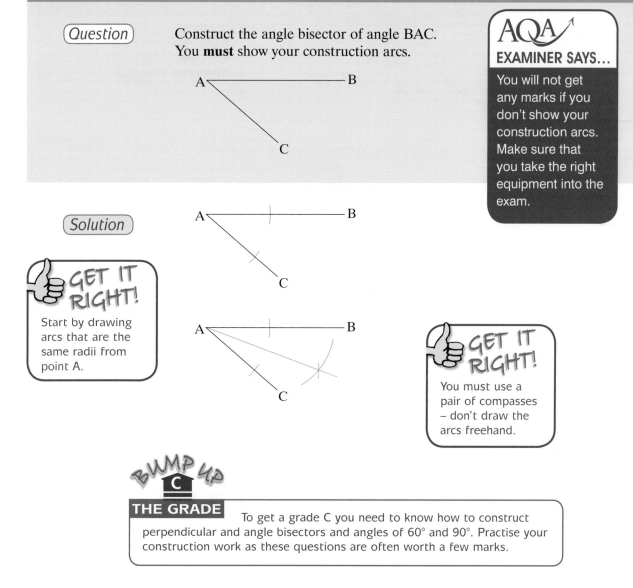

BUMP UP
C
THE GRADE   To get a grade C you need to know how to construct perpendicular and angle bisectors and angles of 60° and 90°. Practise your construction work as these questions are often worth a few marks.

Worked Example Loci

Question

The diagram shows three mobile phone masts A, B and C.

Show on the diagram the region that is less than 6 km from all three masts.

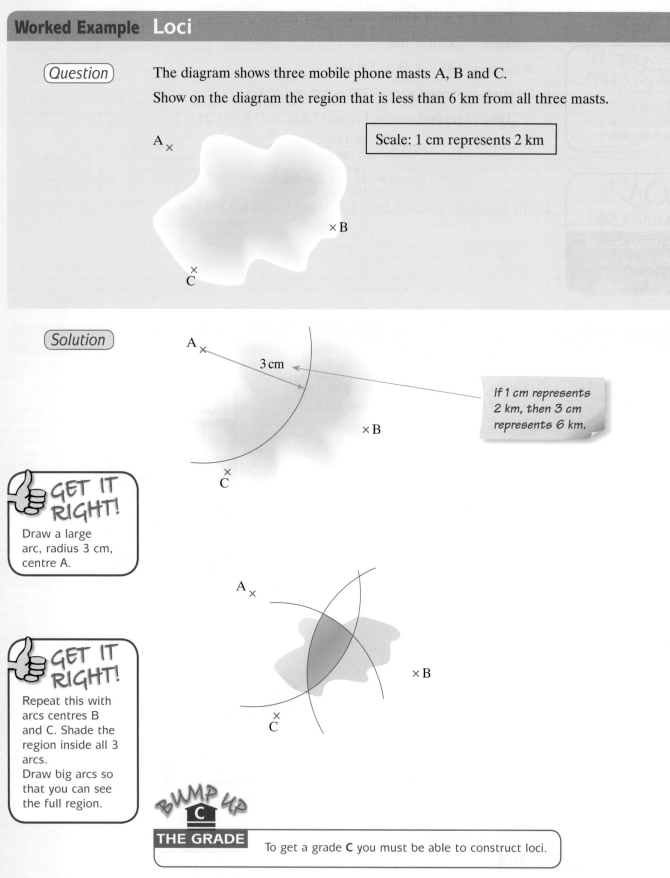

Scale: 1 cm represents 2 km

A×

×B

×
C

Solution

A×

3 cm

×B

×
C

If 1 cm represents
2 km, then 3 cm
represents 6 km.

GET IT RIGHT!

Draw a large
arc, radius 3 cm,
centre A.

A×

×B

×
C

GET IT RIGHT!

Repeat this with
arcs centres B
and C. Shade the
region inside all 3
arcs.
Draw big arcs so
that you can see
the full region.

BUMP UP
C
THE GRADE

To get a grade **C** you must be able to construct loci.

Measures and drawing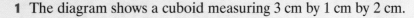

Time Yourself!

Can you complete these questions in **30** minutes?

1 The diagram shows a cuboid measuring 3 cm by 1 cm by 2 cm.

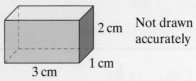

2 cm   Not drawn accurately

3 cm   1 cm

On centimetre square paper draw a net of the cuboid.

2 A packet of crisps weighs 30 g to the nearest gram.

a What are the minimum and maximum weights of the packet?

b The crisps are sold in economy packs of 6.
What are the minimum and maximum weights of an economy pack?

3 a Using a ruler and compasses only, construct an angle of 60°.
Show all your construction arcs.

b Two ships A and B are 6 kilometres apart.
A is due west of B.
Ship A is within 4 kilometres of a lighthouse.
Ship B is within 5 kilometres of the lighthouse.
Using a scale of 1 cm to represent 1 km, accurately construct a diagram showing the position of the ships.
Shade the area in which the lighthouse could be.

4 Draw an 8 cm line.
Draw the locus of all points 2 cm from the line.

5 Ben and Zoë are travelling from North Green to West Park.
Ben rollerblades 6 miles along a winding cycle path at an average speed of 8 mph.
Zoë walks the 4 miles directly at an average speed of 3 mph.
Calculate the difference in time between the two journeys.
Give your answer in minutes.

6 A greyhound runs 100 metres in 8.95 seconds.
Estimate its average speed in kilometres per hour.
You **must** show your working.

7 ABC is a triangle.
AB = 6 cm, AC = 4 cm and angle CAB = 90°.
Construct triangle ABC using a ruler and compasses only.
Show your construction arcs.
Measure and write down the length of BC.

6 Pythagoras' theorem

Key points

◎ **Pythagoras' theorem** – in a **right-angled triangle**, the square of the length of the **hypotenuse** is equal to the sum of the squares of the lengths of the other two sides.

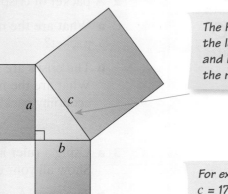

> The hypotenuse is the longest side and is opposite the right angle.

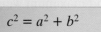
$$c^2 = a^2 + b^2$$

> For example, if $c = 17$ cm, $a = 8$ cm and $b = 15$ cm, then $c^2 = a^2 + b^2$ so this is a right angle.

◎ **Converse of Pythagoras' theorem** – in any triangle if $c^2 = a^2 + b^2$ then the triangle has a **right angle** opposite c.

◎ A **Pythagorean triple** is a set of three integers a, b, c that satisfies $c^2 = a^2 + b^2$

> It helps to learn common Pythagorean triples as they are often used on the non-calculator paper. For example:
>
> 3, 4, 5 ($5^2 = 3^2 + 4^2$),
>
> 5, 12, 13 ($13^2 = 5^2 + 12^2$)

THE GRADE   Using Pythagoras' theorem is a grade **C** topic. For the non-calculator paper you will need to know all the squares of numbers up to 15.

Worked Example **Finding the hypotenuse**

Question   PQR is a right-angled triangle.

PQ = 11 cm and PR = 5.2 cm.

Calculate the length of QR.

Give your answer to an appropriate degree of accuracy.

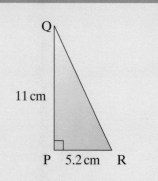

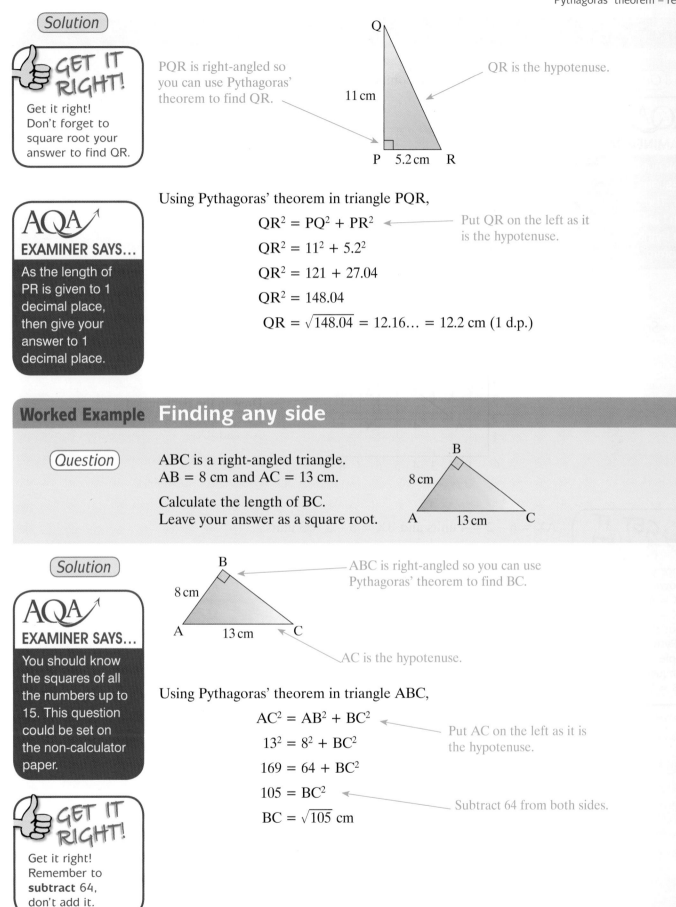

Solution

GET IT RIGHT!

Get it right! Don't forget to square root your answer to find QR.

PQR is right-angled so you can use Pythagoras' theorem to find QR.

QR is the hypotenuse.

Q

11 cm

P   5.2 cm   R

AQA EXAMINER SAYS...

As the length of PR is given to 1 decimal place, then give your answer to 1 decimal place.

Using Pythagoras' theorem in triangle PQR,

$$QR^2 = PQ^2 + PR^2$$

Put QR on the left as it is the hypotenuse.

$$QR^2 = 11^2 + 5.2^2$$

$$QR^2 = 121 + 27.04$$

$$QR^2 = 148.04$$

$$QR = \sqrt{148.04} = 12.16\ldots = 12.2 \text{ cm (1 d.p.)}$$

Worked Example   Finding any side

Question

ABC is a right-angled triangle.
AB = 8 cm and AC = 13 cm.

Calculate the length of BC.
Leave your answer as a square root.

B
8 cm
A   13 cm   C

Solution

B
8 cm
A   13 cm   C

ABC is right-angled so you can use Pythagoras' theorem to find BC.

AC is the hypotenuse.

AQA EXAMINER SAYS...

You should know the squares of all the numbers up to 15. This question could be set on the non-calculator paper.

Using Pythagoras' theorem in triangle ABC,

$$AC^2 = AB^2 + BC^2$$

Put AC on the left as it is the hypotenuse.

$$13^2 = 8^2 + BC^2$$

$$169 = 64 + BC^2$$

$$105 = BC^2$$

Subtract 64 from both sides.

$$BC = \sqrt{105} \text{ cm}$$

GET IT RIGHT!

Get it right! Remember to **subtract** 64, don't add it.

Worked Example Applying Pythagoras' theorem

The diagram shows the points A(2, 2) and B(6, 5).

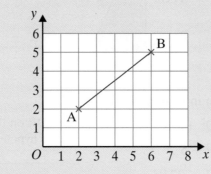

Calculate the distance AB.

Solution

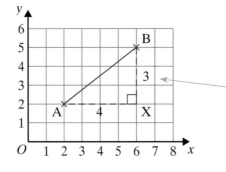

Draw in the right-angled triangle and work out AX and BX.

AX = 6 – 2 = 4 units and BX = 5 – 2 = 3 units

Using Pythagoras' theorem in triangle ABX,

$$AB^2 = AX^2 + BX^2$$

Put AB on the left as it is the hypotenuse.

$$AB^2 = 4^2 + 3^2$$
$$= 16 + 9$$
$$= 25$$
$$AB = \sqrt{25} = 5 \text{ units}$$

Pythogoras' theorem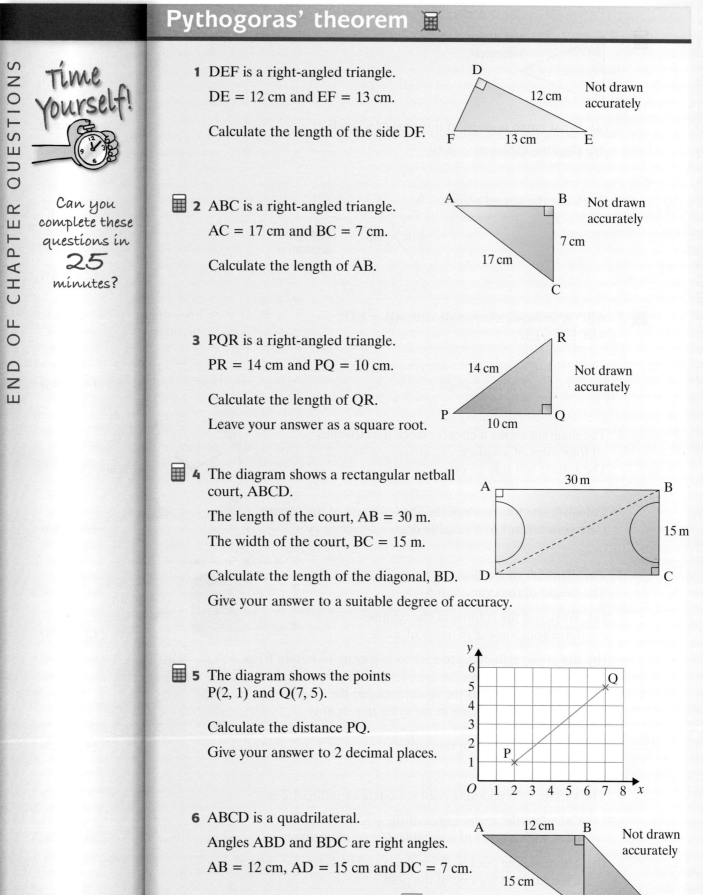

END OF CHAPTER QUESTIONS

Time Yourself!

Can you complete these questions in **25** minutes?

1 DEF is a right-angled triangle.

DE = 12 cm and EF = 13 cm.

Calculate the length of the side DF.

Not drawn accurately

2 ABC is a right-angled triangle.

AC = 17 cm and BC = 7 cm.

Calculate the length of AB.

Not drawn accurately

3 PQR is a right-angled triangle.

PR = 14 cm and PQ = 10 cm.

Calculate the length of QR.

Leave your answer as a square root.

Not drawn accurately

4 The diagram shows a rectangular netball court, ABCD.

The length of the court, AB = 30 m.

The width of the court, BC = 15 m.

Calculate the length of the diagonal, BD.

Give your answer to a suitable degree of accuracy.

5 The diagram shows the points P(2, 1) and Q(7, 5).

Calculate the distance PQ.

Give your answer to 2 decimal places.

6 ABCD is a quadrilateral.

Angles ABD and BDC are right angles.

AB = 12 cm, AD = 15 cm and DC = 7 cm.

Show that the length of BC is $\sqrt{130}$ cm.

Not drawn accurately

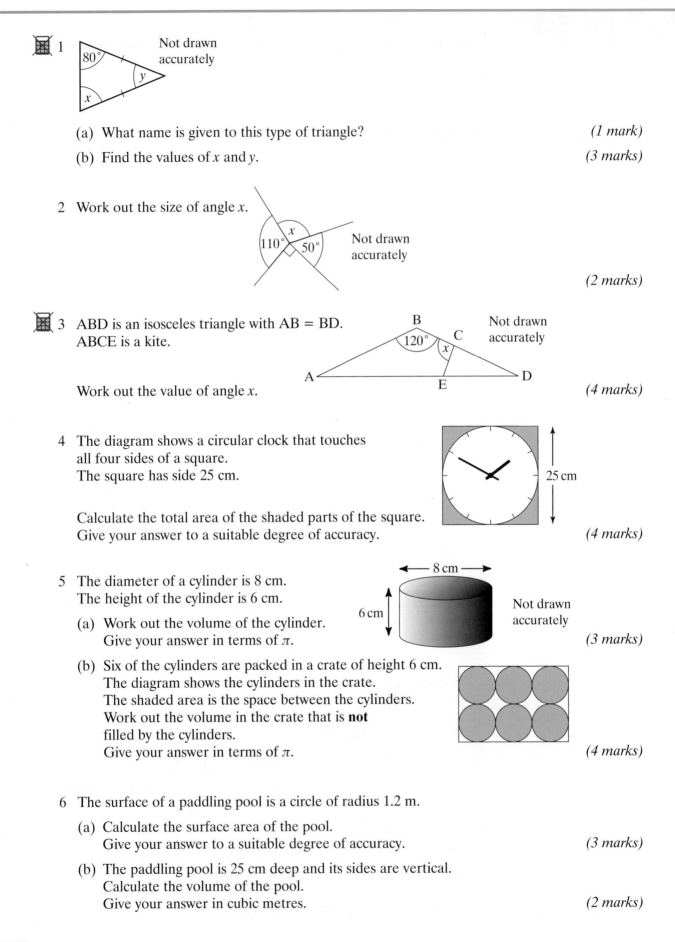

1   Not drawn accurately

(a) What name is given to this type of triangle?   *(1 mark)*

(b) Find the values of x and y.   *(3 marks)*

2   Work out the size of angle x.

Not drawn accurately

(2 marks)

3   ABD is an isosceles triangle with AB = BD.
    ABCE is a kite.

    Not drawn accurately

    Work out the value of angle x.   *(4 marks)*

4   The diagram shows a circular clock that touches
    all four sides of a square.
    The square has side 25 cm.

    Calculate the total area of the shaded parts of the square.
    Give your answer to a suitable degree of accuracy.   *(4 marks)*

5   The diameter of a cylinder is 8 cm.
    The height of the cylinder is 6 cm.

    (a) Work out the volume of the cylinder.
         Give your answer in terms of π.   *(3 marks)*

    (b) Six of the cylinders are packed in a crate of height 6 cm.
         The diagram shows the cylinders in the crate.
         The shaded area is the space between the cylinders.
         Work out the volume in the crate that is **not**
         filled by the cylinders.
         Give your answer in terms of π.   *(4 marks)*

6   The surface of a paddling pool is a circle of radius 1.2 m.

    (a) Calculate the surface area of the pool.
         Give your answer to a suitable degree of accuracy. *(3 marks)*

    (b) The paddling pool is 25 cm deep and its sides are vertical.
         Calculate the volume of the pool.
         Give your answer in cubic metres.   *(2 marks)*

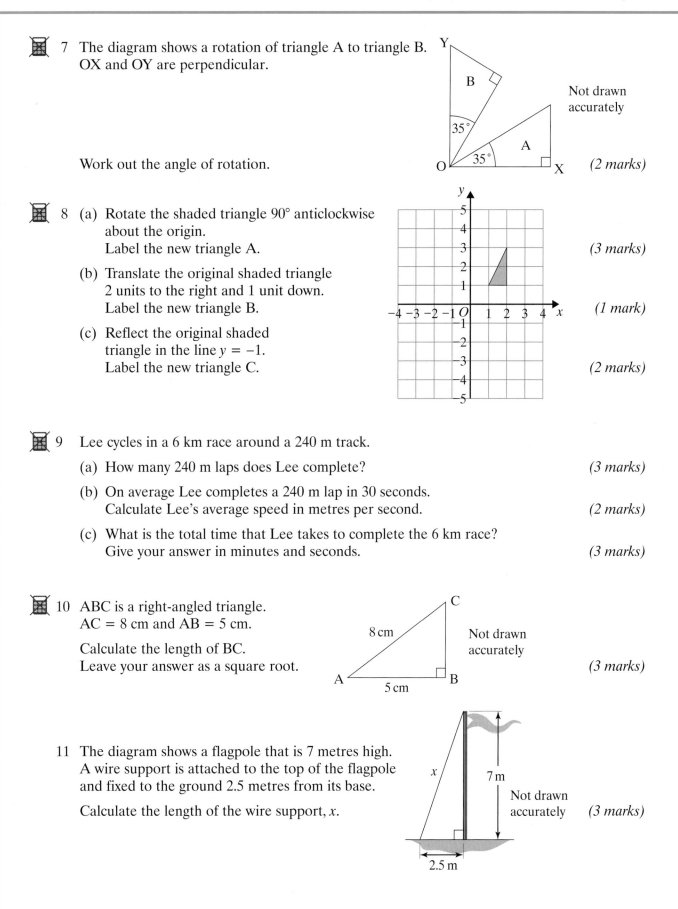

7   The diagram shows a rotation of triangle A to triangle B.
    OX and OY are perpendicular.

    Work out the angle of rotation.   *(2 marks)*

8   (a) Rotate the shaded triangle 90° anticlockwise
         about the origin.
         Label the new triangle A.   *(3 marks)*

    (b) Translate the original shaded triangle
         2 units to the right and 1 unit down.
         Label the new triangle B.   *(1 mark)*

    (c) Reflect the original shaded
         triangle in the line $y = -1$.
         Label the new triangle C.   *(2 marks)*

9   Lee cycles in a 6 km race around a 240 m track.

    (a) How many 240 m laps does Lee complete?   *(3 marks)*

    (b) On average Lee completes a 240 m lap in 30 seconds.
         Calculate Lee's average speed in metres per second.   *(2 marks)*

    (c) What is the total time that Lee takes to complete the 6 km race?
         Give your answer in minutes and seconds.   *(3 marks)*

10 ABC is a right-angled triangle.
    AC = 8 cm and AB = 5 cm.

    Calculate the length of BC.
    Leave your answer as a square root.   *(3 marks)*

11 The diagram shows a flagpole that is 7 metres high.
    A wire support is attached to the top of the flagpole
    and fixed to the ground 2.5 metres from its base.

    Calculate the length of the wire support, x.   *(3 marks)*

1 Adam is drawing a quadrilateral with both of these properties:

| 4 equal sides | | diagonals intersect at right angles |

He draws a square.

(a) Matthew draws a different type of quadrilateral with these properties.
    Draw Matthew's quadrilateral.

The mark is awarded for any rhombus with four equal sides.

The angles should not be 90°, otherwise it is just a rotated square.

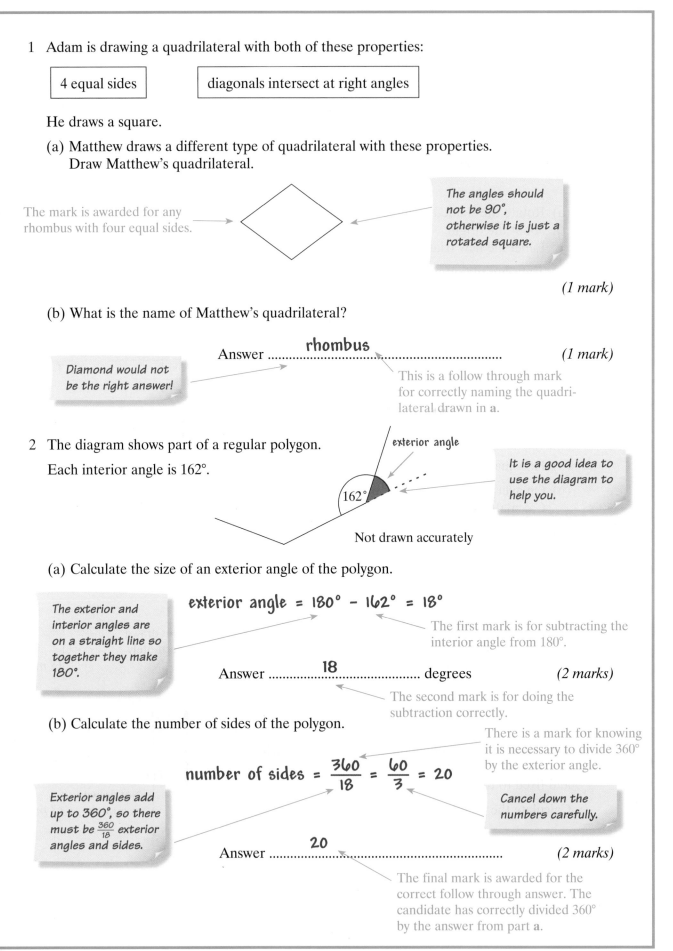

(1 mark)

(b) What is the name of Matthew's quadrilateral?

Answer ....**rhombus**.................................... *(1 mark)*

Diamond would not be the right answer!

This is a follow through mark for correctly naming the quadrilateral drawn in **a**.

2 The diagram shows part of a regular polygon.

Each interior angle is 162°.

162°

exterior angle

It is a good idea to use the diagram to help you.

Not drawn accurately

(a) Calculate the size of an exterior angle of the polygon.

The exterior and interior angles are on a straight line so together they make 180°.

exterior angle = 180° – 162° = 18°

The first mark is for subtracting the interior angle from 180°.

Answer ....**18**.................... degrees *(2 marks)*

The second mark is for doing the subtraction correctly.

(b) Calculate the number of sides of the polygon.

There is a mark for knowing it is necessary to divide 360° by the exterior angle.

number of sides = $\frac{360}{18}$ = $\frac{60}{3}$ = 20

Exterior angles add up to 360°, so there must be $\frac{360}{18}$ exterior angles and sides.

Cancel down the numbers carefully.

Answer ....**20**.................................... *(2 marks)*

The final mark is awarded for the correct follow through answer. The candidate has correctly divided 360° by the answer from part **a**.

3 The diagram shows a triangular prism.

(a) Its shaded cross-section is an isosceles triangle with base 12 cm and perpendicular height 8 cm.

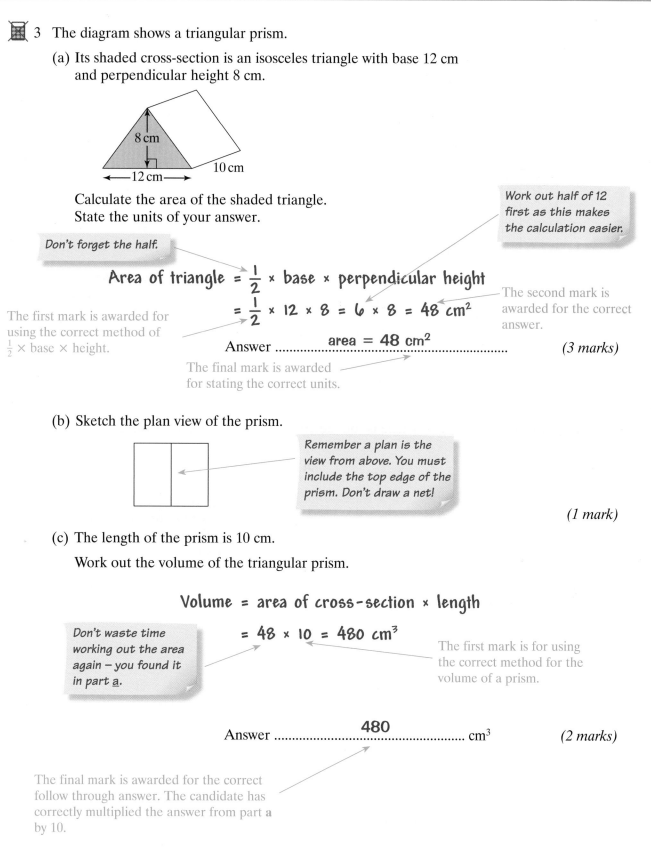

Calculate the area of the shaded triangle.
State the units of your answer.

Work out half of 12 first as this makes the calculation easier.

Don't forget the half.

$$\text{Area of triangle} = \frac{1}{2} \times \text{base} \times \text{perpendicular height}$$

$$= \frac{1}{2} \times 12 \times 8 = 6 \times 8 = 48 \text{ cm}^2$$

The first mark is awarded for using the correct method of $\frac{1}{2} \times$ base $\times$ height.

The second mark is awarded for the correct answer.

Answer ....**area = 48 cm^2**.............. *(3 marks)*

The final mark is awarded for stating the correct units.

(b) Sketch the plan view of the prism.

Remember a plan is the view from above. You must include the top edge of the prism. Don't draw a net!

(1 mark)

(c) The length of the prism is 10 cm.

Work out the volume of the triangular prism.

$$\text{Volume} = \text{area of cross-section} \times \text{length}$$

$$= 48 \times 10 = 480 \text{ cm}^3$$

Don't waste time working out the area again – you found it in part <u>a</u>.

The first mark is for using the correct method for the volume of a prism.

Answer ....**480**................... cm^3 *(2 marks)*

The final mark is awarded for the correct follow through answer. The candidate has correctly multiplied the answer from part **a** by 10.

4

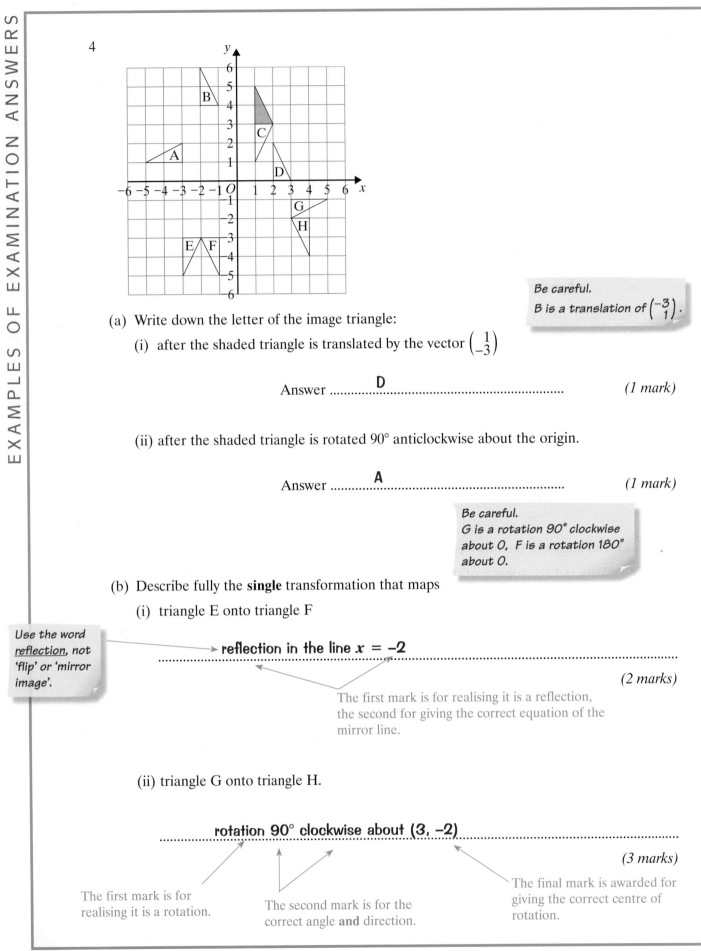

(a) Write down the letter of the image triangle:

(i) after the shaded triangle is translated by the vector $\begin{pmatrix} 1 \\ -3 \end{pmatrix}$

Be careful.
B is a translation of $\begin{pmatrix} -3 \\ 1 \end{pmatrix}$.

Answer ....**D**... *(1 mark)*

(ii) after the shaded triangle is rotated 90° anticlockwise about the origin.

Answer ....**A**... *(1 mark)*

Be careful.
G is a rotation 90° clockwise about O, F is a rotation 180° about O.

(b) Describe fully the **single** transformation that maps

(i) triangle E onto triangle F

Use the word <u>reflection</u>, not 'flip' or 'mirror image'.

→ **reflection in the line $x = -2$**

(2 marks)

The first mark is for realising it is a reflection, the second for giving the correct equation of the mirror line.

(ii) triangle G onto triangle H.

rotation 90° clockwise about (3, –2)

(3 marks)

The first mark is for realising it is a rotation.

The second mark is for the correct angle **and** direction.

The final mark is awarded for giving the correct centre of rotation.

Symbols and sequences

sequence

term

*n*th term

product

simplify

expand

factorise

indices

powers

Key points

◎ A **sequence** is a list of numbers that are connected in some way.

◎ A **term** is a number, a variable, or the product of a number and a variable, for example, n, 3, $4x^2$.

◎ The ***n*th term** is the general term in a sequence.

◎ The **product** of two numbers is the result of multiplying them together.

◎ To **simplify** means to collect like terms (adding or subtracting terms with the same variable).

◎ To **expand** means to remove brackets by multiplying.

◎ To **factorise** means to put brackets back in an expression by finding common factors.

◎ **Powers** (or **indices**) are the small numbers used to show a number or term multiplied by itself, for example, $a^3 = a \times a \times a$.

Worked Example | **Collecting like terms**

Question | Simplify by collecting like terms, $3a + 2a^2 + 4a - 3a^2$.

Solution |
$$3a + 2a^2 + 4a - 3a^2 = 3a + 4a + 2a^2 - 3a^2$$
$$= 7a - a^2$$

GET IT RIGHT!

Be careful with the signs. Make sure you can add and subtract negative numbers. Don't forget: a and a^2 are **not** like terms.

Worked Example | **Multiplying indices**

Question | Simplify $3a^4b^2 \times 2a^3bc$.

b is the same as b^1

Solution |
$$3a^4b^2 \times 2a^3bc = 3 \times a^4 \times b^2 \times 2 \times a^3 \times b \times c$$
$$= 3 \times 2 \times a^4 \times a^3 \times b^2 \times b \times c$$
$$= 6\,a^7b^3\,c \quad \longleftarrow \text{Don't forget to multiply the } 3 \times 2, \text{ but just add the indices.}$$

GET IT RIGHT!

To **multiply** powers of a, just **add** the indices.

Worked Example Dividing indices

Question   Simplify $6a^2b^3 \div 3ab$

Solution $6a^2b^3 \div 3ab = (6 \div 3)(a^2 \div a)(b^3 \div b)$ ⟵ Work out each letter separately.
Don't forget to divide the $6 \div 3$;
but just subtract the indices

$= 2ab^2$ ⟵

GET IT RIGHT!

Get it right!
To **divide** powers
of a, just **subtract**
the indices.

BUMP UP

C

THE GRADE   To get a grade C you must be able to deal with fractions and negative numbers; $\frac{a^3}{a^5}$ means $a^3 \div a^5 = a^{3-5} = a^{-2}$

Worked Example Expanding brackets

Question   Expand and simplify $3(2y - 4) - 3y(4y - 2)$

Solution

×	2y	−4
3	6y	−12

×	4y	−2
−3y	−12y²	+6y

Use the grid method to multiply out brackets.

$= 6y - 12$ $-12y^2 + 6y$

$= 12y - 12y^2 - 12$

GET IT RIGHT!

Make sure you
can multiply
negative numbers.

Worked Example Factorising

Question   Factorise completely $4x^2 - 6xy$

Solution $4x^2 - 6xy = 2x(2x - 3y)$ ⟵

$4x^2 = 2x \times 2x$
$6xy = 2x \times 3y$

BUMP UP

C

THE GRADE   To get a grade **C** you must make sure you take out all the common factors; don't just take out 2 if you could take out $2x$. Always check that you can't factorise the contents of the brackets any further.

AQA

EXAMINER SAYS...

Candidates need
to be able to deal
with one term
being equal to the
common factor,
for example, $6xy + 3y = 3y(2x + 1)$

Worked Example Finding terms in a sequence

Question   Here is a sequence of numbers:

3, 7, 15, 31, ...

The rule for continuing the sequence is: **multiply** by 2 and **add** 1.

What are the next two numbers in the sequence?

Solution   63, 127 ⟵

$31 \times 2 + 1 = 63$
$63 \times 2 + 1 = 127$

Worked Example Finding the *n*th term

Question

Find the *n*th term in this sequence:

5, 8, 11, 14, 17, …

Solution

nth term = difference × n + (first term − difference)

$$= 3n + (5 - 3)$$

$$= 3n + 2$$

Difference = 3
First term = 5

GET IT RIGHT!

Check your rule on any term, for example, when $n = 4$, the 4th term = $3 \times 4 + 2 = 14$

BUMP UP THE GRADE **C**

To get a grade **C** you must make sure you use n in your nth term. Don't write '× 3 + 2' instead of '$3n + 2$'.

Symbols and sequences

Time Yourself!

Can you complete these questions in **10** minutes?

END OF CHAPTER QUESTIONS

1 **a** Factorise completely $3x^2 + 6x^3$

   b Expand and simplify $4x(x^2 + 5) - 3(2x - 4)$

2 Simplify:

   a $3x - 2y - x + 3y$

   b $4x^3y^2 \times 2xy^5$

   c $\dfrac{12x^2}{3x^6}$

3 Patterns are made from rectangles and triangles.

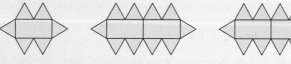

   1st pattern   2nd pattern   3rd pattern

   a How many triangles will there be in the *n*th pattern?

   b How many triangles will there be in the 20th pattern?

2 Equations

Key words

equation

unknown

solution

formula

subject of a formula

inverse operation

integer

inequality

trial and improvement

square root

cube root

decimal places

Key points

◎ An **equation** is a statement showing that two expressions are equal, for example, $2y - 7 = 15$. In this example, the **unknown** is y. The solution is $y = 11$.

◎ A **formula** is an equation that contains two or more variables, for example, $A = \dfrac{bh}{2}$. The **subject of the formula** is A, as the formula tells you how to work out A.

◎ **Inverse operations** undo an operation, for example, -8 is the inverse of $+8$.

◎ An **integer** is a whole number, such as 4, 0 or −3.

◎ An **inequality** shows that two expressions are **not** equal, for example, $3x - 2 > 12$.

◎ **Trial and improvement** is a way of solving equations by making improved guesses.

◎ The **square root** of a number is the number that has to be multiplied by itself (or **squared**) to reach that number, for example, the square root of 16, or $\sqrt{16}$, is 4, as $4 \times 4 = 16$.

◎ **Decimal places** are the digits to the right of a decimal point, for example, 2.34 has two decimal places.

Worked Example Equations

Question Solve the equation $4x - 2 = 6x - 9$

Solution

$$4x - 2 = 6x - 9$$

$$4x - 2 + 9 = 6x - 9 + 9 \quad \longleftarrow \text{Add 9 to both sides}$$

$$4x + 7 = 6x$$

$$4x - 4x + 7 = 6x - 4x \quad \longleftarrow \text{Subtract } 4x \text{ from both sides}$$

$$7 = 2x$$

$$\frac{7}{2} = \frac{2x}{2} \quad \longleftarrow \text{Divide both sides by 2}$$

$$3.5 = x$$

$$x = 3.5$$

The key to solving equations is to use inverse operations to simplify the equation.

GET IT RIGHT!

Aim to get all the unknown terms on one side, and the numbers on the other side.

Worked Example Equations with brackets

Question   Solve the equation $3(2a + 4) = 2a + 4$

Solution

$$3(2a + 4) = 2a + 4$$

$$6a + 12 = 2a + 4 \quad \longleftarrow \text{Multiply out the brackets first}$$

$$6a - 2a + 12 = 2a - 2a + 4 \quad \longleftarrow \text{Subtract } 2a \text{ from both sides}$$

$$4a + 12 = 4$$

$$4a + 12 - 12 = 4 - 12 \quad \longleftarrow \text{Subtract 12 from both sides}$$

$$4a = -8$$

$$\frac{4a}{4} = \frac{-8}{4} \quad \longleftarrow \text{Divide both sides by 4}$$

$$a = -2$$

AQA
EXAMINER SAYS...
Candidates should write their answers as $a = -2$, not just -2

Worked Example Equations with fractions

Question   Solve the equation $\frac{2x}{3} - 5 = 1$

Solution

GET IT RIGHT!

Don't confuse
$\frac{2x}{3} - 5 = 1$
with $\frac{2x - 5}{3} = 1$.
In $\frac{2x - 5}{3} = 1$ the
last operation
is ÷ 3

$$\frac{2x}{3} - 5 = 1$$

$$\frac{2x}{3} - 5 + 5 = 1 + 5 \quad \longleftarrow \text{Add 5 to both sides}$$

$$\frac{2x}{3} = 6$$

$$\frac{2x}{3} \times 3 = 6 \times 3 \quad \longleftarrow \text{Multiply both sides by 3}$$

$$2x = 18$$

$$\frac{2x}{2} = \frac{18}{2} \quad \longleftarrow \text{Divide both sides by 2}$$

$$x = 9$$

Undo the last operation first.
In this case, the last operation was −5

Worked Example Inequalities

Question   Find the smallest integer that satisfies the inequality $2x + 3 > 8 - x$

Solution

Solve inequalities just like equations, using inverse operations.

$$2x + 3 > 8 - x$$

$$2x + x + 3 > 8 - x + x \quad \longleftarrow \text{Add } x \text{ to both sides}$$

$$3x + 3 > 8$$

$$3x + 3 - 3 > 8 - 3 \quad \longleftarrow \text{Subtract 3 from both sides}$$

$$3x > 5$$

$$\frac{3x}{3} > \frac{5}{3} \quad \longleftarrow \text{Divide both sides by 3}$$

$$x > 1\tfrac{2}{3}$$

The smallest integer $> 1\tfrac{2}{3}$ is 2.

AQA
EXAMINER SAYS...
Some candidates forget to give their answer in the right form. If a question asks for an integer, don't leave your answer as a fraction or decimal.

Worked Example Trial and improvement

Question

Use trial and improvement to find a positive solution to the equation
$x^2 - x = 18$. Give your answer to 2 decimal places.

Solution

Trial value	$x^2 - x$	Comment
4	$16 - 4 = 12$	Too low
5	$25 - 5 = 20$	Too high
4.5	$20.25 - 4.5 = 15.75$	Too low
4.7	$22.09 - 4.7 = 17.39$	Too low
4.8	$23.04 - 4.8 = 18.24$	Too high
4.77	$22.7529 - 4.77 = 17.9829$	Too low
4.78	$22.8484 - 4.78 = 18.0684$	Too high
4.775	$22.800625 - 4.775 = 18.025625$	Too high

Once you know that x is between 4.77 and 4.78, try halfway between them (4.775) to see whether the answer is closer to 4.77 or 4.78

$x = 4.77$ (2 d.p.)

AQA EXAMINER SAYS...

Some candidates don't show every step of their working out. It is essential to do so in trial and improvement questions

Worked Example Substituting into a formula

Question

Bob is using the scientific formula, $s = ut + \frac{1}{2}at^2$

Find the value of s if $a = 4$, $t = 8$ and $u = 5$.

Solution

$s = ut + \frac{1}{2}at^2$ ⟵ Start with the formula

$s = (5 \times 8) + (\frac{1}{2} \times 4 \times 8^2)$

$s = 40 + 128$

$s = 168$

Substitute the given values

Remember it's only the 8 that is squared

AQA EXAMINER SAYS...

Many candidates pick up marks for correct working even when they make an error at some stage.

Worked Example   Changing the subject of a formula

Question   Rearrange the formula $A = \dfrac{(a + b)}{2}h$ to make a the subject.

Solution

The method is exactly the same as for solving an equation – use inverse operations.

The formula says:

Start with a; Add b;

Divide by 2;

Multiply by h.

The inverse operations used to undo this are:

Divide by h;

Multiply by 2;

Subtract b.

$$A = \frac{(a + b)}{2}h \quad\longleftarrow\ \text{Start with the formula}$$

$$\frac{A}{h} = \frac{(a + b)}{2}h \div h \quad\longleftarrow\ \text{Divide both sides by } h$$

$$\frac{A}{h} = \frac{(a + b)}{2}$$

$$\frac{2A}{h} = \frac{2(a + b)}{2} \quad\longleftarrow\ \text{Multiply both sides by 2}$$

$$\frac{2A}{h} = a + b$$

$$\frac{2A}{h} - b = a + b - b \quad\longleftarrow\ \text{Subtract } b \text{ from both sides}$$

$$\frac{2A}{h} - b = a$$

$$a = \frac{2A}{h} - b$$

Equations

END OF CHAPTER QUESTIONS

Time Yourself!

Can you complete these questions in **15** minutes?

1 a Use the formula $v = u + at$ to find the value of v when $u = -6$, $a = 2.4$ and $t = 1.9$

  b Solve the equations:

   i $5x - 3 = 2 + 3x$ ii $\dfrac{12 - x}{2} = 6.5$ iii $3(2x + 1) = 7 - 5x$

2 If $x = 2$ and $y = -5$, find the value of:

  a $3x + 2y$ b $\dfrac{x - 2y}{3}$

3 Make a the subject of the formula $x = an - b$

4 Use trial and improvement to find the positive solution to the equation $x^2 - x = 9$. Give your answer to 1 decimal place.

5 Solve the inequality $5x - 11 < 2x + 1$

3 Coordinates and graphs

coordinates

origin

axis (pl. axes)

gradient

intercept

midpoint

equidistant

conversion graph

speed

Key points

- **Coordinates** identify a point.

- The **origin** is the point $(0, 0)$.

- The **axes** are the lines used to locate points.

- The **gradient** tells you is how steep a line is, calculated as

$$\frac{\text{change in vertical distance}}{\text{change in horizontal distance}}$$

- The **intercept** is the y-coordinate of the point where a line crosses the y-axis.

- The **midpoint** is the middle point of a line.

- If **A** and **B** are equidistant from **C**, the distance **AC** = the distance **BC**.

- A **conversion graph** is used to convert one unit into another, for example, pounds to kilograms.

- **Speed** is calculated as $\dfrac{\text{distance}}{\text{time}}$.

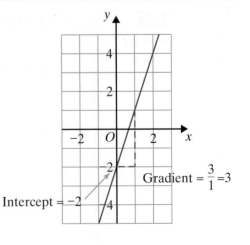

y-axis

y-coordinate 2 -------- P (3, 2)

O   3   x-axis

x-coordinate

Origin

Worked Example Equations and straight lines

(Question) Draw the line that has the equation $y = 3x - 2$

(Solution)

GET IT RIGHT!

Don't forget that lines such as $x = 4$ are vertical (all the points on it have an x-coordinate of 4), and lines such as $y = -2$ are horizontal.

The graph of a straight line can always be written in the form $y = mx + c$, where m is the gradient and c is the intercept on the y-axis.

So $y = 3x - 2$ crosses the y-axis at -2 with a gradient of 3.

You can check your graph by making sure the coordinates fit the equation. For example, the graph passes through $(2, 4)$, so when $x = 2$, $y = 4$:

When $x = 2$, $y = 3x - 2 = 3 \times 2 - 2 = 4$

Gradient $= \dfrac{3}{1} = 3$

Intercept $= -2$

Worked Example Problems about coordinates

(Question) Where does the line $y = 7$ cross the line $y = 2x + 1$?

(Solution) $y = 7$ tells us that the y-coordinate is 7.

$y = 2x + 1$

$7 = 2x + 1$ ⟵——— Substitute y = 7

$6 = 2x$ ⟵——— Subtract 1 from both sides of the equation

$x = 3$ ⟵——— Divide both sides by 2

The lines cross at (3, 7).

AQA
EXAMINER SAYS...
Candidates often solve the equation and then stop. Remember to answer the question and give the coordinates.

Worked Example Problems about gradients

(Question) Write down the equation of the line parallel to $y = 2x - 5$, which passes through the point (0, 1).

(Solution) Parallel lines have the same gradient, so the gradient is 2.

(0, 1) is on the y-axis, so the intercept is 1.

The equations is $y = 2x + 1$

Worked Example The midpoint of a line segment

(Question) Write down the coordinates of the point halfway between (2, –5) and (7, 1).

(Solution) The midpoint is
$\left(\dfrac{2 + 7}{2}, \dfrac{-5 + 1}{2} \right)$
or (4.5, –2)

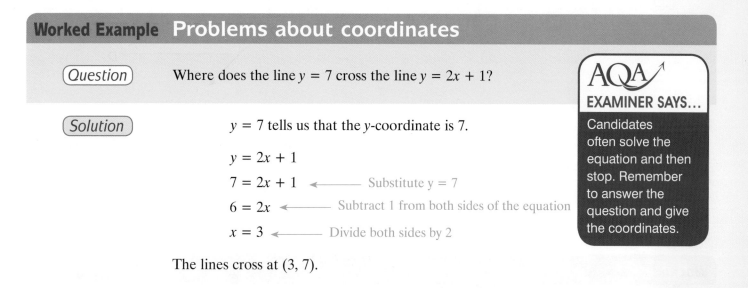

The midpoint of the line joining (a, b) and (c, d) is

$$\left(\frac{a + c}{2}, \frac{b + d}{2} \right)$$

The mean of the x-coordinates

The mean of the y-coordinates

AQA
EXAMINER SAYS...
Candidates should always draw a rough sketch to make sure their answer seems sensible.

Worked Example Coordinates in three dimensions

Question   In the diagram, C is the point (4, 8, 5).

Write down the coordinates of A, B and D.

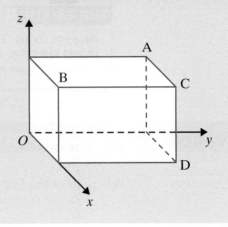

Solution

B, C and D have the same *x*-coordinate.

A, C and D have the same *y*-coordinate.

A, B and C have the same *z*-coordinate.

C	A	B	D
(4, 8, 5)	(0, ,)	(4, ,)	(4, ,)
(4, 8, 5)	(0, 8,)	(4, 0,)	(4, 8,)
(4, 8, 5)	(0, 8, 5)	(4, 0, 5)	(4, 8, 0)

Worked Example Real-life graphs

Question   The graph shows Bob's journey from home to his grandmother's house.

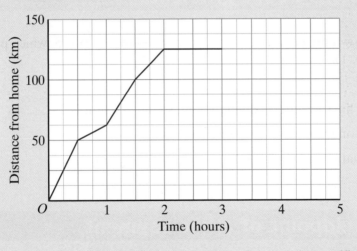

a After how long did Bob get stuck in slow-moving traffic?

b What was his average speed during the first half hour of his journey?

c Bob arrived at his grandmother's house after 2 hours. What was his average speed over the whole journey?

d Bob stopped at his grandmother's house for an hour. He then went home at a speed of 100 km/h. Draw his return journey on the graph.

Solution

a After half an hour Bob got stuck in slow-moving traffic – the gradient shows he went slowly.

b Bob travelled 50 km in half an hour, which is 100 km/h.

c Bob travelled 125 km in 2 hours.

d Speed = $\dfrac{\text{distance}}{\text{time}}$ = 62.5 km/h

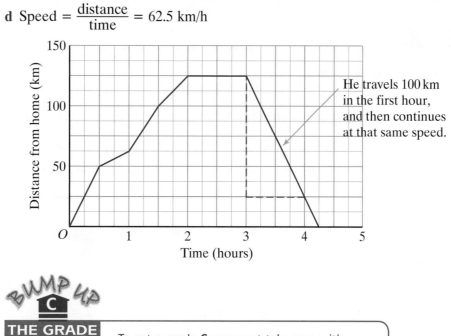

He travels 100 km in the first hour, and then continues at that same speed.

BUMP UP

C

THE GRADE To get a grade **C** you must take care with units. Some travel graphs are in minutes, and there are 60 minutes in an hour, so don't write 45 minutes as 0.45 hours.

Coordinates and graphs

END OF CHAPTER QUESTIONS

Time Yourself!

Can you complete these questions in **10** minutes?

1 Draw a grid with the *x*-axis from 0 to 5 and the *y*-axis from 0 to 15. On the grid, draw the graph of $y = 2x + 5$

2 A is the point (0, 4) and B is the point (2, 0).

a Find the midpoint of AB.

b What is the gradient of the line through AB?

c What is the equation of the line through AB?

d What is the equation of the line parallel to AB that passes through the point (0, −1)?

3 The diagram shows a pyramid. The point A has coordinates (3, 2, 5) and is directly over the centre of the rectangular base. What are the coordinates of the point B?

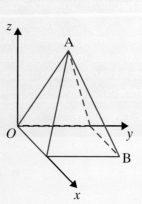

4

Quadratic functions

Key points

⊚ **Quadratic functions** have an x^2-term, but no higher power of x.

⊚ Graphs of quadratic functions are always ∪-shaped (when the x^2-term is positive) or ∩-shaped (when the x^2-term is negative).

Worked Example **Graphs of quadratic functions**

Question

a Draw the graph of $y = 2x^2 - 3x - 2$ for values of x from -3 to 4.

b Use the graph to find the solutions of $2x^2 - 3x - 2 = 0$.

Solution

a

x	-3	-2	-1	0	1	2	3	4
$2x^2$	18	8	2	0	2	8	18	32
$-3x$	9	6	3	0	-3	-6	-9	-12
-2	-2	-2	-2	-2	-2	-2	-2	-2
$y = 2x^2 - 3x - 2$	25	12	3	-2	-3	0	7	18

Don't forget that $2x^2$ means $2 \times x \times x$.

Don't work out $2x$ and then square the answer

The table breaks the equation into three parts, $2x^2$, $-3x$ and -2. These three parts are added together to calculate y.

AQA

EXAMINER SAYS…

Many candidates misuse their calculators to work out x^2 when x is negative. Some calculators give $2 \times -3^2 = -18$ instead of the correct answer of 18. Candidates must remember to use brackets with negative numbers:

$2 \times (-3)^2 = 18$.

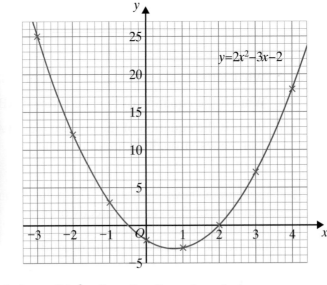

$y = 2x^2 - 3x - 2$

Note the scales: each small square on the x-axis is 0.2, but on the y-axis is 1

b The solutions of $2x^2 - 3x - 2 = 0$ are found where $y = 0$, which is the x-axis. From the graph, these values are $x = -0.5$ and $x = 2$.

Worked Example **Recognising equations of quadratic functions**

(Question) Match the equations with their graphs.

a $y = x^2 - 5$

b $y = 5 - x^2$

c $y = 5 - x$

(Solution) **a** $y = x^2 - 5$ is a quadratic (x^2) graph, and so is ∪-shaped. The −5 in the equation tells you that it crosses the y-axis at −5. So it is the **black** graph.

b $y = 5 - x^2$ is quadratic, but the x^2 is negative, so it is ∩-shaped. The 5 in the equation indicates that it crosses the y-axis at 5. So it is the blue graph.

c $y = 5 - x$ is linear (a straight line). The −x indicates it has a gradient of −1, so it slopes downwards from left to right. The 5 shows that it crosses the y-axis at 5, so it is the **red** graph.

Quadratic functions

END OF CHAPTER QUESTIONS

Time Yourself!

Can you complete these questions in **15** minutes?

1 a Complete the table of values for $y = x^2 - 2x$ for values of x from −3 to 3.

x	−3	−2	−1	0	1	2	3
x^2							
$-2x$							
y							

b Draw the graph of $y = x^2 - 2x$, for values of x from −3 to 3.

c Write down the x-coordinates of the points where the graph crosses the line $y = 2$.

2 a Draw the graph of $y = x^2$ for values of x from −5 to 5.

b Draw the line $y = x + 1$ on the same axes.

c Write down the x-coordinates of the points where the curve and the line cross.

d Write down the quadratic equation whose solutions are the answers to part **c**.

3 Draw a sketch of the graphs of the following equations.

a $y = x + 2$ **b** $y = 2 - x$ **c** $y = x^2 + 2$ **d** $y = 2 - x^2$

5 Proof

sum

product

consecutive numbers

counter-example

proof

Key points

- The **sum** of numbers is the result of adding them together.
- The **difference** of two numbers is the result of subtracting one number from the other.
- The **product** is the result of multiplying numbers.
- **Consecutive numbers** are next to each other, for example, 3, 4, 5.
- A **counter-example** is an example that disproves a statement.
- A **proof** is a logical explanation that shows something must be true.

Worked Example Proof

Question

a and *b* are both less than 10.
Is it true that $a \times b$ must be less than 100?

Solution

If $a = -12$ and $b = -9$, then *a* and *b* are both less than 10, but $a \times b = 108$
This is a counter-example, so the statement is **false**.

GET IT RIGHT!

Remember how to multiply negative numbers.

THE GRADE To get a grade **C** you should remember that
negative numbers, zero, and fractions (or decimals) often provide
counter examples for statements that appear to be true.

Worked Example Algebraic proofs

Question

Brian writes down the number   74

He reverses the digits to get   47

He subtracts to get 74 – 47 =   27, which is 3 × 9

He repeats this with different two-digit starting numbers.

He notices that the answer is always a multiple of 9.

Prove that this is always true.

(Solution)   74 has a value of 7 tens + 4 units.

So a two-digit number, written as ab, has a value of $10a + b$

Reversing it gives ba, which has a value of $10b + a$.

Subtracting gives $(10a + b) - (10b + a)$

$$= 10a + b - 10b - a$$

$$= 9a - 9b$$

$$= 9(a - b), \text{ which is a multiple of } 9$$

AQA

EXAMINER SAYS…

Candidates need to recognise that a two-digit number ab represents $10a + b$, and that even numbers can be represented by $2n$, and odd numbers by $2n + 1$.

Worked Example Odd, even and multiples

(Question)   Prove that the sum of four consecutive numbers is always even but is never a multiple of 4.

(Solution)   Let the consecutive numbers be $a, a + 1, a + 2$ and $a + 3$.

$$a + a + 1 + a + 2 + a + 3 = 4a + 6$$

$$= 2(2a + 3), \text{ which is a multiple of 2 and so is even.}$$

But $4a + 6 = 4a + 4 + 2$

$$= 4(a + 1) + 2, \text{ which is 2 more than a multiple of 4}$$

Proof

Time Yourself!

Can you complete these questions in **15** minutes?

END OF CHAPTER QUESTIONS

1 Show that the sum of three consecutive numbers is always a multiple of 3.

2 Is it true that all prime numbers can be written as the sum of two consecutive numbers?

3 Look at the grid below.

1	2	3	4	5
6	7	8	9	10
11	12	13	14	15
16	17	18	19	20
21	22	23	24	25
26	27	28	29	30
31	32	33	34	35
36	37	38	39	40

The four grey cells in the 2 × 2 grid add up to 44, which is a multiple of 4.

Prove that in **any** of the 2 × 2 blocks of cells, the sum of the four numbers is a multiple of 4.

1 Simplify $(2g^2h^5)(3g^3h)$.   *(2 marks)*

2 Factorise completely $6ab - 4b$.   *(1 mark)*

3 This question is about the sequence

   $3, 7, 11, 15, 19, \ldots$

   Write an expression for the nth term of the sequence.   *(2 marks)*

4 Solve the equation $2(3a - 1) = a + 8$.   *(2 marks)*

5 Make r the subject of the formula $p = 4r + t$.   *(2 marks)*

6 The distance, s metres, travelled by a football is given by the formula

   $s = ut + \frac{1}{2}at^2$

   Calculate s when $a = 9$, $t = 4.2$ and $u = -2$.   *(3 marks)*

7 The equation $x^3 + x = 20$ has a solution between 2 and 3.

   Use trial and improvement to find the solution correct to 2 decimal places.   *(3 marks)*

8 A is the point $(4, -1)$ and B is the point $(2, 4)$. Find the coordinates of C,
   the midpoint of AB.   *(2 marks)*

9 A hexagon is placed against two walls, so that A
   and B rest against one wall, and C touches another
   wall. The coordinates of A are $(6, 0, 5)$. The
   coordinates of E are $(2, 5, 0)$. The coordinates
   of C are $(0, 2.5, 2.5)$. Find the coordinates
   of B, D and F.   *(6 marks)*

10 (a) Draw the graph of $y = x^2 + x$, taking values of x from -4 to 4.   *(3 marks)*

    (b) On the same axes, draw the graph of $y = 5$.   *(1 mark)*

    (c) Write down the coordinates of the points of intersection of the two graphs.   *(2 marks)*

11 P is a prime number. Q is an even number.

    State whether each of the following is always odd or always even or could be either even or
    odd.

    (a) $Q(P + 1)$ *(1 mark)*

    (b) $P - Q$ *(1 mark)*

1 Factorise $x^2 + 7x$.

> x is a common factor:
> $x^2 = x \times x$, $7x = x \times 7$

Answer .... $x(x + 7)$ .... *(1 mark)*

2 Expand $x^3(5 + 3x)$.

> Using the grid method

×	5	+3x
x^3	$5x^3$	$+3x^4$

> $5x^3 + 3x^4$ can't be simplified.

Answer .... $5x^3 + 3x^4$ .... *(2 marks)*

One mark is given for each correct term.

3 Solve the inequality $6y > 2y + 10$.

$$6y - 2y > 2y - 2y + 10$$

$$4y > 10$$ ← First mark for subtracting $2y$ from each side

$$4y \div 4 > 10 \div 4$$ Second mark for dividing both sides by 4

Answer .... $y > 2.5$ .... *(2 marks)*

4 Solve the equation $\frac{x}{2} - 1 = \frac{x}{4} + 2$

> Multiply everything by 4 (the LCM of the denominators)

Either

$$\frac{x}{2} - 1 + 1 = \frac{x}{4} + 2 + 1$$

$$\frac{x}{2} = \frac{x}{4} + 3$$ ← First mark

$$\frac{x}{2} - \frac{x}{4} = \frac{x}{4} - \frac{x}{4} + 3$$

$$\frac{2x}{4} - \frac{x}{4} = 3$$ Second mark

$$\frac{x}{4} = 3$$

Answer .... $x = 12$ ← Third mark

Or

$$\frac{4x}{2} - 4 = \frac{4x}{4} + 8$$

$$2x - 4 = x + 8$$

$$2x - 4 + 4 = x + 8 + 4$$

$$2x = x + 12$$

$$2x - x = x - x + 12$$

$$x = 12$$

Answer .... $x = 12$ *(3 marks)*

5 Find the equation of a straight line parallel to $y = 2x - 1$, which passes through the point (0, 3)

The candidate recognises the intercept as 3

The intercept on the y-axis is 3. The gradient is 2.

Answer .... $y = 2x + 3$ .... *(2 marks)*

One mark is awarded for stating $2x$, and one mark for $+ 3$

The candidate recognises the gradient as 2

ANSWERS

Section 1: . Handling data

1 Collecting data

1 **a** The number of people at a rugby match is quantitative as it can be counted.
   b How many tins of beans a shop sells is quantitative as it can be counted.
   c The flavour of the beans is qualitative.
   d The time it takes to travel from London to Manchester is quantitative as it can be measured.

2 **a** The number of votes for a party at a local election is discrete.
   b The number of beans in a tin is discrete.
   c The weight of a tin of beans is continuous as weight is a measurement and is always continuous.
   d The time taken to complete this chapter is continuous as time is a measurement and is always continuous.

3 **a** There are 3 + 2 + 1 + 2 = 8 red cars
   b There are 2 + 4 + 3 + 0 = 9 Vauxhall cars
   c Number of black cars = 7
   Total number of cars = 35
   Percentage = $\frac{7}{35} \times 100 = 20\%$

4 **a** Assign a number to each student in the population and generate random numbers to choose a random sample of 50 students.
   b Assign each student in the population a number and generate a random number to start, then sample every nth student, where n = population size divided by 50.

2 Statistical measures

1 **a** mean = 40 ÷ 8 = 5
   median = 5
   There are two modes: 3 and 5
   range = 10 − 1 = 9
   b mean = 200 ÷ 8 = 25
   median = 25
   There are two modes: 23 and 25
   range = 30 − 21 = 9
   c mean = 17 ÷ 9 = 1.89 (3 s.f.)
   median = 2
   there are two modes: 2 and 3
   range = 3 − 0 = 3

2 Mean = (19 + 10 + 17 + 18 + 19 + 18 + 17 + 22 + 7 + 13 + 18) ÷ 11
   = 178 ÷ 11 = 16.2 (3 s.f.)
   For the median arrange the numbers in order:
   7 10 13 17 17 18 18 18 19 19 22
   Median = 18
   Mode = 18
   Range = 22 − 7 = 15

3 Median = 8th item = 4
   Mode = 4
   Mean = 64 ÷ 15 = 4.27 (3 s.f.)
   Range = 8 − 2 = 6

4 **a** Modal class = £10–£20
   b Construct a table for the mean:

Pocket money £	Number of students (f)	Midpoint (x)	Midpoint (×) frequency (fx)
0–10	7	5	7 × 5 = 35
10–20	9	15	9 × 15 =135
20–30	6	25	6 × 25 =150
30–40	3	35	3 × 35 =105
Total	$\sum f$ =25		$\sum fx$ = **425**

Mean = 425 ÷ 25 = £17

3 Representing data

1 **a** Pictogram

Fruit juice	Frequency
Vanilla	♈ ♈ ♈ ♈ ♈ ♈ ♈ ♈
Strawberry	♈ ♈ ♈ ♈ ♈
Raspberry	♈ ♈ ♈
Mango	♈ ◝
Other	◝

Key ♈ = 2 ice cream creams

b Bar chart

Ice cream sales

(Bar chart: Frequency on y-axis from 0 to 18, ice creams on x-axis: Vanilla 16, Strawberry 10, Raspberry 6, Mango 3, Other 1)

Ice creams

c Pie chart
The pie chart needs to be drawn to represent 36 ice creams.
There are 360° in a full circle so each ice cream will be shown by 360° ÷ 36 =10°

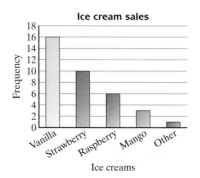

Ice cream sales

Ice cream	Frequency	Angle
Vanilla	16	16 × 10° = 160°
Strawberry	10	10 × 10° = 100°
Raspberry	6	26 × 10° = 60°
Mango	3	3 × 10° = 30°
Other	1	1 × 10° = 10°
	36	360°

2 a The heights of shrubs in a garden centre

Stem (whole number)	Leaf
1	55 11 82 93
2	31 79 44 51 66
3	23 04

Key: 2|51 represents 2.51 m

b To find the median it is important to have the data in numerical order.
This can be done by creating an ordered stem-and-leaf diagram.

The heights of shrubs in a garden centre

Stem (whole number)	Leaf
1	11 55 82 93
2	31 44 51 66 79
3	04 23

> Here the leaves (units) are all arranged numerically.

The median = 2.44 m

3 a $72 \times 4 = 288$ students were included in the survey (as 90° represents 72 students)
  b $\frac{1}{3} \times 288 = 96$ students chose 'Non-fiction' books
  c The remaining sectors represent
  $288 - (72 + 96) = 120$ students
  'Romance' books account for $\frac{2}{3}$ of this total, that is,
  $\frac{2}{3} \times 120 = 80$ students

1 Scatter graphs

1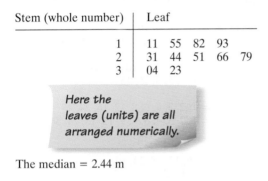

a Weak negative correlation

b Moderate negative correlation

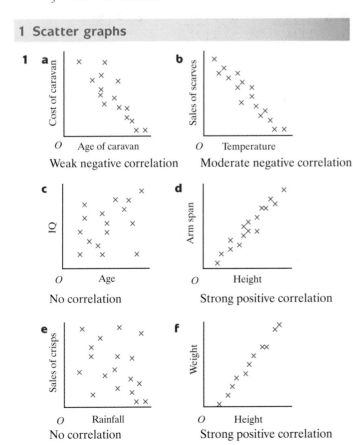

c No correlation

d Strong positive correlation

e No correlation

f Strong positive correlation

2 a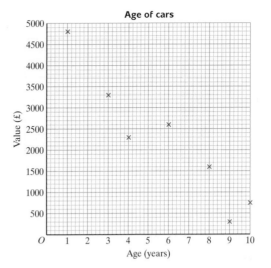

Age of cars

b The scatter graph shows moderate negative correlation.

3 a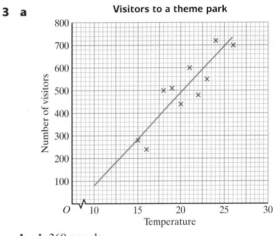

Visitors to a theme park

b i 360 people
  ii 12°C
  c The temperature is more likely to be inaccurate because it goes outside the given range of values.

5 Probability

1 a Number of blue marbles $= 0.3 \times 30 = 9$
  b Number of red marbles $= \frac{1}{5} \times 30 = 6$
  c There are 15 marbles left that must be green so probability $= \frac{15}{30} = \frac{1}{2}$

2 The table shows the frequency distribution after drawing a card from a pack 60 times.
  a Relative frequency of getting a club $= \frac{11}{60}$
  b Relative frequency of getting a red card $= \frac{27}{60}$ (there are 27 red cards)
  c Theoretical probability of getting a spade $= \frac{1}{4}$

3 a

Second dice

First dice		1	2	3	4	5	6
	1	2	3	4	5	6	7
	2	3	4	5	6	7	8
	3	4	5	6	7	8	9
	4	5	6	7	8	9	10
	5	6	7	8	9	10	11
	6	7	8	9	10	11	12

a i $\frac{3}{36} = \frac{1}{12}$
ii $\frac{6}{36} = \frac{1}{6}$
iii 0

b The modal score is 7 (most frequent)

4 $1 - 0.85 = 0.15$

5 P(Orange) $= 1 - (0.35 + \frac{1}{5} + 0.1) = 0.35$

Section 1: Handling data

ANSWERS TO EXAMINATION STYLE QUESTIONS

1 Answer column has 0, 7, 18, 12, 12, 5
Mean $= \frac{54}{30} = 1.8$ magazines

2 Midpoints are 45, 55, 65, 75
Mean $= 3480 \div 60 = 58$ mph

3 Midpoints are 3, 5, 7, 9, 11, 13
Mean $= 228 \div 30 = 7.6$ min (3 s.f.)

4 Any three of the following:
Not ordered
7 omitted in stem
69 only appears once (should be twice) or
only 14 data items.

5 a Your line of best fit should be a straight line with positive gradient. It should cover the lengths 40–58 cm, beginning on or below the point (40, 2.5). Your line should travel through or between (50, 3.7) and (52, 3.5), and also through or between (56, 4.5) and (57, 4.2).
b Approximately 4.2 kg

6

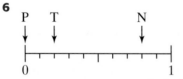

7 a Relative frequency $= \frac{5}{20}, \frac{7}{20}, \frac{8}{20}$ or 0.25, 0.35, 0,4
b Yes, as the spinner has been spun more times

8 a Price index in 1995 = 100
When the price is 20% more than this the price index is 120, which was in 2000.
b Price index in 1995 = 100
Price index in 2006 = 152
Therefore, increase in price = 52%

Section 2: Number

Make sure you can also do these on a calculator.

1 Integers and rounding

1 a $\frac{6-9}{3} = \frac{-3}{3} = -1$ **b** $\frac{-1+9}{-2} = \frac{8}{-2} = -4$

c $\frac{-5 \times 8}{-4} = \frac{-40}{-4} = 10$ **d** $\frac{-7 \times -6}{4} = \frac{42}{4} = 10.5$

2 a 15, 35 **b** 40, 72 **c** 15 **d** 35 **e** 54 **f** 37

3 a Factors of 24 = 1, 2, 3, 4, 6, 8, 12, 24
Factors of 32 = 1, 2, 4, 8, 16, 32
Common factors are 1, 2, 4, 8
b Factors of 36 = 1, 2, 3, 4, 6, 9, 12, 18, 36
Factors of 48 = 1, 2, 3, 4, 6, 8, 12, 16, 24, 48
Common factors are 1, 2, 3, 4, 6, 12

4 a Factors of 20 = 1, 2, 4, 5, 10, 20
Factors of 36 = 1, 2, 3, 4, 6, 9, 12, 18, 36
HCF = 4
b Factors of 30 = 1, 2, 3, 5, 6, 10, 15, 30
Factors of 45 = 1, 3, 5, 9, 15, 45
HCF = 15
c Factors of 48 = 1, 2, 3, 4, 6, 8, 12, 16, 24, 48
Factors of 72 = 1, 2, 3, 4, 6, 8, 9, 12, 18, 24, 36, 72
HCF = 24

5 a Multiples of 6 = 6, 12, 18, 24, 30, …
Multiples of 15 = 15, 30, …
LCM = 30
b Multiples of 12 = 12, 24, 36, 48, 60,…
Multiples of 20 = 20, 40, 60, …
LCM = 60
c Multiples of 20 = 20, 40, 60, 80, 100, 120, 140, 160, 180, …
Multiples of 36 = 36, 72, 108, 144, 180, …
LCM = 180

6 a

2	28
2	14
7	7
	1

$28 = 2^2 \times 7$

b

2	54
3	27
3	9
3	3
	1

$54 = 2 \times 3^3$

c

2	90
3	45
3	15
5	5
	1

$90 = 2 \times 3^2 \times 5$

d

2	120
2	60
2	30
3	15
5	5
	1

$120 = 2^3 \times 3 \times 5$

e

2	250
5	125
5	25
5	5
	1

$250 = 2 \times 5^3$

f

3	567
3	189
3	63
3	21
7	7
	1

$567 = 3^4 \times 7$

7 a $\frac{1}{8}$ **b** 3 **c** $\frac{1}{0.4} = \frac{10}{4} = \frac{5}{2} = 2\frac{1}{2}$ or 2.5

d $\frac{8}{5} = 1\frac{3}{5}$ or 1.6   **e** $-\frac{1}{4}$ or -0.25

8 a 8 000 000 **b** 8072 **c i** 7830 **ii** 8000

9 a $\frac{206 \times 4.91}{1.92^2} \approx \frac{200 \times 5}{2 \times 2} = \frac{1000}{4} = 250$

b $\frac{5.74 + 3.12}{0.879} \approx \frac{6+3}{0.9} = \frac{9}{0.9} = \frac{90}{9} = 10$

c $\frac{83.9}{19.3 \times 0.163} \approx \frac{80}{20 \times 0.2} = \frac{80}{4} = 20$

d $\frac{86.71 - 32.04}{41.56 + 79.52} \approx \frac{90 - 30}{40 + 80} = \frac{60}{120} = 0.5$

10 Lower bound = 795 litres
Upper bound = 805 litres

2 Decimals and fractions

Make sure you can also do these on a calculator.

1 a 36.25 **b** 8.82 **c** 2.128 **d** $\dfrac{8.5}{0.05} = \dfrac{850}{5} = 170$

e $48.52 + 6 \div 10 = 48.52 + 0.6 = 49.12$
f $3.2 \times 100 - 8.4 \times 10 = 320 - 84 = 236$
g $0.4 \times (5 - 1.3) = 0.4 \times 3.7 = 1.48$
h $3 \times 1.4 + 0.5 \times 0.3 = 4.2 + 0.15 = 4.35$

2 a $£27.50 + 4 \times £12.25 = £76.50$
b $£125.50 - £27.50 = £98$
Number of **extra** days = $£98 \div £12.25 = 8$
Total number of days = 9

3 a 0.025, 0.06, 0.125, 0.28, 0.4
b $0.4 = \dfrac{4}{10} = \dfrac{2}{5}$ $0.06 = \dfrac{6}{100} = \dfrac{3}{50}$ $0.28 = \dfrac{28}{100} = \dfrac{7}{25}$
$0.025 = \dfrac{25}{1000} = \dfrac{1}{40}$ $0.125 = \dfrac{125}{1000} = \dfrac{5}{40} = \dfrac{1}{8}$

4 a $\dfrac{3}{4} = 0.75$ **b** $\dfrac{4}{5} = 0.8$ **c** $\dfrac{2}{3} = 0.\dot{6}$

d $\dfrac{11}{20} = 0.55$ **e** $\dfrac{7}{11} = 0.\dot{6}\dot{3}$ $11\overline{)7\,.\,{}^7 0 {}^4 0 {}^7 0\ ...}$ $0\,.\,6\,3\,6\,...$

5 a $\dfrac{2}{6} = \dfrac{1}{3}$ **b** $\dfrac{9}{12} = \dfrac{3}{4}$ **c** $\dfrac{8}{12} = \dfrac{2}{3}$ **d** $\dfrac{4}{10} = \dfrac{2}{5}$

6 a $\dfrac{3}{5}$ of 30 kg = $30 \div 5 \times 3 = 18$ kg
b $\dfrac{5}{8}$ of £140 = $£140 \div 8 \times 5 = £17.50 \times 5 = £87.50$

7 $\dfrac{9}{15}$ and $\dfrac{24}{40}$

8 a $\dfrac{7}{10} = \dfrac{21}{30}$ $\dfrac{3}{5} = \dfrac{18}{30}$ $\dfrac{5}{6} = \dfrac{25}{30}$ $\dfrac{2}{3} = \dfrac{20}{30}$
In order: $\dfrac{3}{5}, \dfrac{2}{3}, \dfrac{7}{10}, \dfrac{5}{6}$ **b** $\dfrac{7}{10}$ is nearest to $\dfrac{2}{3}$

9 a $\dfrac{1}{3} + \dfrac{1}{6} = \dfrac{2}{6} + \dfrac{1}{6} = \dfrac{3}{6} = \dfrac{1}{2}$

Check you can also do these on a calculator.

b $\dfrac{7}{8} - \dfrac{3}{5} = \dfrac{35}{40} - \dfrac{24}{40} = \dfrac{11}{40}$
c $\dfrac{2}{7} \times \dfrac{5}{6} = \dfrac{5}{21}$
d $\dfrac{3}{4} \div \dfrac{9}{20} = \dfrac{3}{4} \times \dfrac{20}{9} = \dfrac{5}{3} = 1\dfrac{2}{3}$
e $3\dfrac{2}{3} + 1\dfrac{4}{5} = 4 + \dfrac{10}{15} + \dfrac{12}{15} = 4 + \dfrac{22}{15} = 4 + 1\dfrac{7}{15} = 5\dfrac{7}{15}$
f $4\dfrac{1}{2} - 2\dfrac{5}{8} = 2 + \dfrac{4}{8} - \dfrac{5}{8} = 1 + \dfrac{12}{8} - \dfrac{5}{8} = 1\dfrac{7}{8}$

10 a Karen: $\dfrac{2}{5}$ of £240 = $£240 \div 5 \times 2 = £48 \times 2 = £96$
b Liam: $\dfrac{3}{8}$ of £240 = $£240 \div 8 \times 3 = £30 \times 3 = £90$
c Nick: $£240 - £186 = £54$
d £54 out of £240 = $\dfrac{54}{240} = \dfrac{9}{40}$
Alternative method for (d):
$\dfrac{2}{5} + \dfrac{3}{8} = \dfrac{16}{40} + \dfrac{15}{40} = \dfrac{31}{40}$
Nick's share (the rest) = $\dfrac{9}{40}$

In many questions there are alternative methods.

3 Percentages

1 a $25\% = \dfrac{25}{100} = \dfrac{1}{4}$ **b** $60\% = \dfrac{60}{100} = \dfrac{3}{5}$
c $45\% = \dfrac{45}{100} = \dfrac{9}{20}$ **d** $32\% = \dfrac{32}{100} = \dfrac{8}{25}$

2 a $75\% = 0.75$ **b** $7\% = 0.07$
c $30\% = 0.3$ **d** $55\% = 0.55$

3 a 3% **b** 70% **c** 42% **d** 2.5%
e $\dfrac{3}{4} \times 100 = 75\%$ **f** $\dfrac{9}{10} \times 100 = 90\%$
g $\dfrac{4}{5} \times 100 = 80\%$ **h** $\dfrac{13}{20} \times 100 = 65\%$
i $\dfrac{7}{8} \times 100 = \dfrac{175}{2} = 87.5\%$

4 a $\dfrac{4}{10} = 40\%$ **b** $\dfrac{2}{8} = \dfrac{1}{4} = 25\%$ **c** $\dfrac{6}{20} = \dfrac{3}{10} = 30\%$

5 $\dfrac{5}{8} = 0.625$ and $65\% = 0.65$ In order of size: $\dfrac{5}{8}$, 65%, 0.7

6 a 10% of 650 m = 65 m
40% of 650 m = $65 \times 4 = 260$ m
b 10% of £64 = £6.40
5% of £64 = £3.20
35% of £64 = $£6.40 \times 3 + £3.20 = £22.40$
c 25% of 156 kg = $\dfrac{1}{4}$ of 156 = $156 \div 4 = 39$ kg
75% of 156 kg = $39 \times 3 = 117$ kg

7 25% of £580 = $580 \div 4 = £145$ Cost = $£580 - £145 = £435$

8 a 10% of £340 = £34   5% of £340 = £17
2.5% of £340 = £8.50
Total = $£340 + £59.50 = £399.50$
b 10% of £16 = £1.60   5% of £16 = £0.80
2.5% of £16 = £0.40   Total = $£16 + £2.80 = £18.80$
c 10% of £26.80 = £2.68   5% of £26.80 = £1.34
2.5% of £26.80 = £0.67
Total = $£26.80 + £4.69 = £31.49$

9 4.5% of £360 = $£360 \div 100 \times 4.5 = £16.20$
Pay after rise = $£360 + £16.20 = £376.20$

10 a $\dfrac{56}{80} \times 100 = 70\%$ **b** $\dfrac{15}{60} \times 100 = 25\%$ fail   75% pass

11 Loss = $£480 - £320 = £160$ % loss = $\dfrac{160}{480} \times 100 = 33\dfrac{1}{3}\%$

12 Buying price for 1 pen = 40p, Selling price = 65p
Profit = 25p   Percentage profit = $\dfrac{25}{40} \times 100 = 62\dfrac{1}{2}\%$

4 Ratio and proportion

1 a $35:28 = 5:4$
b $48:64 = 3:4$
c $200:2500 = 2:25$
d $75:120:300 = 5:8:20$

There are often different ways to simplify a ratio. for example in 1b dividing by 8 first $48:64 = 6:8 = 3:4$ or halving $48:64 = 24:32 = 12:16 = 6:8 = 3:4$

2 a $1:1.6$
b $1:2.5$
c $5:200 = 1:40$
d $250:1800 = 1:7.2$

3 a $£200 \div 5 = £40$
1st share = £40, 2nd share = $£40 \times 4 = £160$
b $£60 \div 8 = £7.50$
1st share = $£7.50 \times 5 = £37.50$
2nd share = $£7.50 \times 3 = £22.50$
c $£7500 \div 25 = £300$
1st share = $£300 \times 7 = £2100$
2nd share = $£300 \times 8 = £2400$
3rd share = $£300 \times 10 = £3000$
d $£48 \div 10 = £4.80$
1st share = $£4.80 \times 5 = £24$
2nd share = $£4.80 \times 3 = £14.40$
3rd share = $£4.80 \times 2 = £9.60$

Remember to check the total.

4 $180° \div 6 = 30°$ Largest angle = $3 \times 30° = 90°$

5 **a** 9 parts = 36 kg so 1 part = 4 kg. Tin = 2 parts = 8 kg
b 1 part = 82.5 kg ÷ 11 = 7.5 kg. Tin = 2 parts = 15 kg

6 R : S : T = 12 : 18 : 20 = 6 : 9 : 10
1 part = £140 ÷ 25 = £5.60
Ruth gets £5.60 × 6 = £33.60, Sue gets £5.60 × 9 = £50.40,
Tom gets = £5.60 × 10 = £56

7 1 kg will do 72 ÷ 3 = 24 washes.
5 kg will do 24 × 5 = 120 washes

8 For 18 students, cost = £45
For 6 students, cost = £45 ÷ 3 = £15
For 24 students, cost = £15 × 4 = £60

Cost is £15 for 6 students.
Cost is £75 for 6 × 5 = 30 students.
(or £15 for 6 students means £30 for 12 students,
and £60 for 24 students giving £75 for 30 students)

Class	Number of students	Total cost
A	18	£45
B	24	**£60**
C	**30**	£75

You could use cost for
1 student = £45 ÷ 18 =
£2.50 but the arithmetic
is much harder.

9 **a** Table B
(the relationship
is $y = 5x$)

b i

ii The graph shows that y is
proportional to x because
it is a straight line through O.

5 Indices

Make sure you can also do these on a calculator.

1 **a** 9 (3^2), 16 (4^2), 49 (7^2), 64 (8^2)
b 8 (2^3), 27 (3^3), 64 (4^3)

2 **a** 6 × 6 = 36   **b** $\sqrt{9}$ = 3   **c** $\sqrt{100}$ = 10
d $\sqrt[3]{125}$ = 5   **e** $-5^2 = -25$ **f** $(-5)^2 = 25$
g $(-5)^3 = -125$ **h** $(-5)^4 = 625$

3 **a** **i** 15 is just less than 16, so $\sqrt{15} \approx 3.9$
ii 38 is a little more than 36, so $\sqrt{38} \approx 6.2$
iii 76 is nearer 81 than 64, so $\sqrt{76} \approx 8.7$
iv 150 is nearer 144 than 169, so $\sqrt{150} \approx 12.2$
v 200 is a little more than 196, so $\sqrt{200} \approx 14.1$
b **i** $\sqrt{15}$ = 3.87 (2 d.p.)
ii $\sqrt{38}$ = 6.16 (2 d.p.)
iii $\sqrt{76}$ = 8.72 (2 d.p.)
iv $\sqrt{150}$ = 12.25 (2 d.p.)
v $\sqrt{200}$ = 14.14 (2 d.p.)

4 **a** **i** $2^6 = 2 \times 2 \times 2 \times 2 \times 2 \times 2 = 64$
ii $3^4 = 3 \times 3 \times 3 \times 3 = 81$
iii $7^3 = 7 \times 7 \times 7 = 343$
iv $5^4 = 5 \times 5 \times 5 \times 5 = 625$
v $10^5 = 10 \times 10 \times 10 \times 10 \times 10 = 100\,000$
b **i-v** As for answer **4a**

Make sure you can also do these on a calculator.

5 **a** $2^3 \times 3^4 = 2 \times 2 \times 2 \times 3 \times 3 \times 3 \times 3 = 8 \times 81 = 648$
b $5^2 \times 2^5 = 5 \times 5 \times 2 \times 2 \times 2 \times 2 \times 2 = 800$
c $\dfrac{4^2}{2^3} = \dfrac{4 \times 4}{2 \times 2 \times 2} = 2$ Pairing 2s and 5s gives an easy way to calculate this: $10 \times 10 \times 2 \times 2 \times 2$
d $\dfrac{7^2}{10^3} = \dfrac{7 \times 7}{10 \times 10 \times 10} = \dfrac{49}{1000}$ or 0.049
e $\dfrac{1}{0.2^2} = \dfrac{1}{0.2 \times 0.2} = \dfrac{1}{0.04} = \dfrac{100}{4} = 25$

6 **a** 3^4 **b** $7^5 \times 7^4 = 7^{5+4} = 7^9$
c $7^5 \div 7^4 = 7^{5-4} = 7^1 = 7$ **d** $(7^5)^4 = 7^{5 \times 4} = 7^{20}$
e $\dfrac{5^3}{5^7} = 5^{3-7} = 5^{-4}$ 5^{-4} is equal to $\dfrac{1}{5^4}$

7 **a** $\dfrac{6^4 \times 6}{6^2} = \dfrac{6^5}{6^2} = 6^3$ **b** $(5^2 \times 5^4)^3 = (5^6)^3 = 5^{18}$
c $\dfrac{4^7}{4^2 \times 4^3} = \dfrac{4^7}{4^5} = 4^2$
d $\left(\dfrac{2^3}{2}\right)^5 = (2^2)^5 = 2^{10}$

8 Length of each side = $\sqrt{6000}$ = 77.459.... = 77.5 cm (1 d.p.)

9 **a** $\sqrt[3]{2197}$ = 13
b **i** $\dfrac{1}{\sqrt[3]{2197}}$ = 0.07692307692   **ii** 0.077 (2 s.f.)

10 Length of each side = $\sqrt[3]{2.5}$ = 1.36 metres (2 d.p.)

Section 2: Number
ANSWERS TO EXAMINATION STYLE QUESTIONS

1 **a** $56 = 2^3 \times 7$
b Multiples of 56 = 56, 112, 168, ...
Multiples of 84 = 84, 168, ...
LCM = 168

2 **a** $p + q$ is an odd number   **b** pq is an even number

3 Milk is delivered after 2, 4, 6, 8, 10, 12, 14, 16, 18, 20, 22, 24,
26, 28, ... days
Butter is delivered after 4, 8, 12, 16, 20, 24, 28, ... days
Eggs are delivered after 7, 14, 21, 28, ... days.
The deliveries next arrive on the same day after 28 days.

4 $\dfrac{78.32}{0.186 \times 2.19} \approx \dfrac{80}{0.2 \times 2} = \dfrac{80}{0.4} = \dfrac{800}{4} = 200$

5 **a** $\dfrac{7.8 - 2.9}{12.58 - 14.49} = \dfrac{4.9}{-1.91} = -2.565445026$
b -2.6 (2 s.f.)

6 **a** £24.50   **b** £25.49

7 **a** 320   **b** $\dfrac{400}{0.8} = \dfrac{4000}{8} = 500$
c 9345 ÷ 35 = 267 so 9345 ÷ 3.5 = 2670

Treat $ like £, that is, give answers to the nearest cent.

8 **a** 600 × 1.49 = 894 euros
b 586 ÷ 0.81 = $723.46 (2 d.p.)

9 Sunita earns 15 × £6 = £90
Sunita saves $\frac{2}{5}$ of £90 = £90 ÷ 5 × 2
= £18 × 2 = £36 per week
Number of weeks it takes her = £180 ÷ £36 = 5 weeks

10 $4\frac{3}{4} + 2\frac{2}{3} = 6 + \frac{9}{12} + \frac{8}{12} = 6 + \frac{17}{12} = 6 + 1\frac{5}{12} = 7\frac{5}{12}$ metres

11 a 10% of £9.50 = £0.95 or 95p
  b 1% of £2430 = £2430 ÷ 100 = £24.30
  c 25% of £8.40 = £2.10
    75% of £8.40 = £6.30

12 15% of £540 = 540 ÷ 100 × 15 = £81

13 10% of £30 = £3   so 80% of £30 = £3 × 8 = £24
   $\frac{1}{5}$ of £35 = £7   so $\frac{3}{5}$ of £35 = £7 × 3 = £21
                 80% of £30 is larger.

14 10% of £760 = £76 5% of £760 = £38
   15% of £760 = £76 + £38 = £114

15 a 10% of £78 = £7.80
     5% of £78 = £3.90
     2.5% of £78 = £1.95
     Price including VAT = £78 + £13.65 = £91.65
  b $\frac{20}{32} = \frac{5}{8}$ $\frac{5}{8} \times 100 = 62.5\%$

16 a 6.5% of £18 600 = 18 600 ÷ 100 × 6.5 = £1209
     Pay after rise = £1209 + £18 600 = £19 809
  b Pay rise = £20 670 − £19 500 = £1170
     Percentage pay rise = 1170 ÷ 19 500 = 0.06 = 6%

17 For 500 ml of fruit drink you need 140 ml lemonade
   For 250 ml you need 70 ml of lemonade (dividing by 2)
   For 750 ml Robert needs 210 ml (multiplying by 3 or adding)

18 a £360 ÷ 4 = £90
     Avril's share = £90, Brian's share = 3 × £90 = £270
  b Brian's share = $\frac{3}{4}$ = 75%

19 Milk : dark = 18 : 12 = 3 : 2   40 ÷ 5 = 8
   Number of dark chocolates = 2 × 8 = 16

20 a **i** 23   **ii** 27   **b** $2^5 \times 3^2 = 32 \times 9 = 288$

21 The cube root of any number less than 1 is larger than itself,
   for example, $\sqrt[3]{0.1} = 0.464...$ is larger than 0.1

22 a $3^8 \div 3^5 = 3^3 = 27$ **b** $xy^2 = 63 = 7 \times 9 = 7 \times 3^2$
                           so $x = 7$ and $y = 3$

23 a $2^8 = 256$ **b** 0.4
  c **i** $6.4^2 + 3.18^3 + 0.95^4 = 73.93193825$ **ii** 70 (1s.f.)

Section 3: Shape, space and measures

1 Angles

1 a Reflex   **b** Right   **c** Acute   **d** Obtuse

2 $x + 54° + 90° = 180°$ (angles in a triangle)
      $x + 144° = 180°$
           $x = 180° − 144° = 36°$

3. a C   **b** B   **c** Yes, Yes, No.

4 a $x = 30°$ (alternate angles)
  b $y = 70°$ (opposite angles)
  c $z + y + 30° = 180°$ (angles in a triangle)
     $z + 70° + 30° = 180°$
        $z + 100° = 180°$
           $z = 180° − 100° = 80°$

5 a sum of interior angles of a pentagon
     $= (5 − 2) \times 180° = 3 \times 180° = 540°$
     angle C $= \frac{540°}{5} = 108°$ (interior angle of regular
                                  pentagon)
     triangle BCD is isosceles (BC = CD, sides of regular
                                  pentagon)
     angle CBD = angle CDB = x (base angles of isosceles
                                  triangle)
     $2x + 108° = 180°$ (angles in a triangle)
        $2x = 180° − 108° = 72°$
         $x = \frac{72°}{2} = 36°$
  b exterior angle, $y = \frac{360°}{5} = 72°$

2 Perimeter, area and volume

1 a cuboid   **b** cylinder   **c** triangular prism
  d (rectangular-based) pyramid

2 a Area = 10.4 × 2.2 + (8.5 − 2.2) × 2.2 = 36.74 cm²
    or   = 10.4 × 8.5 − (8.5 − 2.2) × 4.1 × 2 = 36.74 cm²
  b Perimeter = large semicircle + small semicircle + line
            $= \frac{\pi \times 12}{2} + \frac{\pi \times 6}{2} + 6 = 9\pi + 6$ cm
            $= 34.3$ cm (1 d.p.)

3 The matchbox must look like:

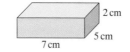

   Volume = length × width × height = 7 × 5 × 2 = 70 cm³

4 Area $= \frac{\pi r^2}{2} = \frac{\pi \times 2.5^2}{2} = 9.81... = 9.8$ m² (1 d.p.)

5 a Width of frame $= \frac{30 − 20}{2} = 5$ cm
     Height of frame = 25 + 5 + 5 = 35 cm
                 $x = 35$ cm
  b Area of frame and mirror = 30 × 35 = 1050 cm²
     Area of mirror = 20 × 25 = 500 cm²
     Area of frame = 1050 − 500 = 550 cm²

6 a Volume = area of cross-section × length
          = $\frac{1}{2}$ × base × height × length
          = $\frac{1}{2}$ × 6 × 8 × 3 = 72 cm³
  b Surface area = 2 × triangle area + area of base + area
                    of back + area of slope
          = 2 × $\frac{1}{2}$ × 6 × 8 + 6 × 3 + 8 × 3 + 10 × 3
          = 120 cm²

7 Surface area = area of top + area of curved face + area of
               base
          = $\pi r^2 + \pi dh + \pi r^2$
          = $\pi \times 3.6^2 + \pi \times (3.6 \times 2) \times 4.2 + \pi \times 3.6^2$
          = 176.43... = 176 cm² (3 s.f.)

8 a Area of trapezium = $\frac{1}{2}(a + b)h$
               = $\frac{1}{2}(1.8 + 2.5) \times 2 = 4.3$ m²
  b Volume = area of cross-section × length
        = 4.3 × 3 = 12.9 m³

3 Reflections and rotations

1 a i 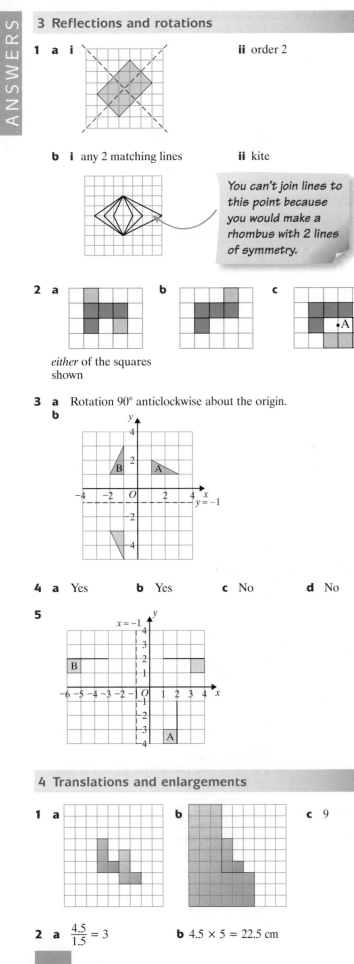   **ii** order 2

b i any 2 matching lines   **ii** kite

You can't join lines to this point because you would make a rhombus with 2 lines of symmetry.

2 a   **b**   **c** •A

either of the squares shown

3 a Rotation 90° anticlockwise about the origin.

b

4 a Yes   **b** Yes   **c** No   **d** No

5

4 Translations and enlargements

1 a   **b**   **c** 9

2 a $\frac{4.5}{1.5} = 3$ **b** $4.5 \times 5 = 22.5$ cm

3 a $240 \times 40 = 9600$ mm $= 960$ cm $= 9.6$ m

b 44 m = 4400 cm
4400 ÷ 40 = 110 cm

4 a 8:12 = 2:3

b 6:*a* = 2:3, *a* = 9 cm

c *b*:6 = 2:3, *b* = 4 cm

d Perimeter of X = 8 + 4 + 4 + 6 = 22 cm
Perimeter of Y = 12 + 6 + 6 + 9 = 33 cm
Ratio = 22:33 = 2:3
(or you might remember that because perimeter is a length it will be the same ratio as for the sides)

5 a   **b** $\begin{pmatrix} -3 \\ 1 \end{pmatrix}$

c

5 Measures and drawing

1 Net of cuboid, for example:

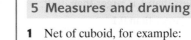
3 cm
2 cm
1 cm
2 cm
2 cm
1 cm

2 a minimum = 29.5 g
maximum = 30.5 g

b minimum = 6 × 29.5 = 177 g
maximum = 6 × 30.5 = 183 g

3 a Draw a straight line.
With point of compasses on P draw a large arc to cut line at Q.
With same radius and point of compasses on Q
draw arc to cut other arc at R.
Join PR.
Angle RPQ = 60°

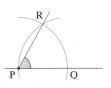

b Draw a line AB = 6 cm.
With point of compasses
on A, radius 4 cm draw
an arc.
With point of compasses
on B, radius 5 cm draw
an arc.
Shade area where arcs
cross.

A |← 4 cm →|
|← 5 cm →| B

4 A straight line 2 cm above
and below original line
joined at each end
by a semicircle,
radius 2 cm.

← 8 cm → 2 cm

5 time = $\dfrac{\text{distance}}{\text{speed}}$
Ben's time = $\frac{6}{8} = \frac{3}{4}$ h = 45 min
Zoë's time = $\frac{4}{3} = 1\frac{1}{3}$ h = 80 min
Difference = 80 – 45 = 35 min

6 8.95 s ≈ 9 s
speed = $\dfrac{\text{distance}}{\text{time}}$
speed = $\frac{100}{9}$ m/s
= $\frac{100}{9} \times 60 \times 60$ metres/ hour
= $\dfrac{100 \times 60 \times 60}{9 \times 1000}$ km/h
= 40 km/h

7 Draw a long line and mark off AB = 6 cm.
Draw two arcs, centre A, equal radius, either side of A.
Centre on each arc, draw arcs that cross above A.
Join this point to A.
Measure 4 cm up this line and
mark off C.
Join BC and measure.
BC = 7.2 cm

C

4 cm

A   6 cm   B

6 Pythagoras' theorem

1 $EF^2 = DF^2 + DE^2$
$13^2 = DF^2 + 12^2$
$169 = DF^2 + 144$
$25 = DF^2$
$DF = \sqrt{25} = 5$ cm   (or recognise a 5, 12, 13 Pythagorean triple)

2 $AC^2 = AB^2 + BC^2$
$17^2 = AB^2 + 7^2$
$289 = AB^2 + 49$
$240 = AB^2$
$AB = \sqrt{240} = 15.49\ldots = 15.5$ cm (1 d.p.)

3 $PR^2 = PQ^2 + RQ^2$
$14^2 = 10^2 + RQ^2$
$196 = 100 + RQ^2$
$96 = RQ^2$
$RQ = \sqrt{96}$ cm

4 In triangle BCD, DC = 30 m, BC = 15 m.
$BD^2 = DC^2 + BC^2$
$BD^2 = 30^2 + 15^2$
$BD^2 = 900 + 225$
$BD^2 = 1125$
$BD = \sqrt{1125} = 33.5\ldots = 34$ m (nearest metre)

5 PX = 7 − 2 = 5 units,
QX = 5 − 1 = 4 units.
$PQ^2 = PX^2 + QX^2$
$PQ^2 = 5^2 + 4^2$
$PQ^2 = 25 + 16$
$PQ^2 = 41$
$PQ = \sqrt{41} = 6.403\ldots = 6.40$ units
(2 d.p.)

6 In triangle ABD
$AD^2 = AB^2 + BD^2$
$15^2 = 12^2 + BD^2$
$225 = 144 + BD^2$
$81 = BD^2$
$BD = \sqrt{81} = 9$ cm (or recognise a 9, 12, 15 Pythagorean triple)
In triangle BDC
$BC^2 = BD^2 + DC^2$
$BC^2 = 9^2 + 7^2$
$BC^2 = 81 + 49$
$BC^2 = 130$
$BC = \sqrt{130}$ cm

Section 3: Shape, space and measures

ANSWERS TO EXAMINATION STYLE QUESTIONS

1 a Isosceles
b $x = 80°$ (base angles of isosceles triangle)
$y + x + 80° = 180°$ (angles in a triangle)
$y + 80° + 80° = 180°$
$y + 160° = 180°$
$y = 180° − 160° = 20°$

2 $x + 50° + 90° + 110° = 360°$ (angles at a point)
$x + 250° = 360°$
$x = 360° − 250° = 110°$

3 angle BAE + angle BDA + 120° = 180° (angles in a triangle)
angle BAE + angle BDA = 180° − 120° = 60°
angle BAE = angle BDA (base angles of isosceles triangle)
angle BAE = $\frac{60°}{2} = 30°$
angle AEC = angle ABC = 120° (kite)
$x + 120° + 120° + 30° = 360°$ (angles in a quadrilateral)
$x + 270° = 360°$
$x = 360° − 270° = 90°$

4 Area of square = 25 × 25 = 625 cm²
Radius of clock = $\frac{25}{2}$ = 12.5 cm
Area of clock = $\pi r^2 = \pi \times 12.5^2 = 490.87\ldots$ cm²
Shaded area = 625 − 490.87 = 134.13 cm² = 134 cm²
(Give your answer to the nearest whole number because the number you were given was to that accuracy.)

5 a Volume = $\pi r^2 h = \pi \times 4^2 \times 6 = 96\pi$ cm³
b Volume of crate = diameter of 3 cylinders × diameter of 2 cylinders × height
= 24 × 16 × 6 = 2304 cm³
Volume of 6 cylinders = 96π × 6 = 576π cm³
Volume of space = 2304 − 576π cm³

6 a Surface area $= \pi r^2 = \pi \times 1.2^2 = 4.52\ldots = 4.5$ m^2 (1 d.p.)
  b Height $= 25$ cm $= 0.25$ m
    Volume $= 4.52\ldots \times 0.25 = 1.13.. = 1.1$ m^3 (1 d.p.)

7 Angle of rotation $= 90° - 35° = 55°$ anticlockwise.

8

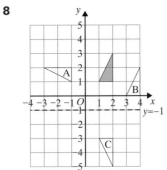

9 a 6 km $= 6000$ m
    number of laps $= 6000 \div 240 = 25$ laps
  b speed $= \dfrac{\text{distance}}{\text{time}} = \dfrac{240}{30} = 8$ m/s
  c 6 km $= 6000$ m
    time $= \dfrac{\text{distance}}{\text{speed}} = \dfrac{6000}{8} = 750$ s
    750 s $= 750 \div 60 = 12.5$ minutes $= 12$ min 30 s

10 $AC^2 = AB^2 + BC^2$
    $8^2 = 5^2 + BC^2$
    $64 = 25 + BC^2$
    $39 = BC^2$
    $BC = \sqrt{39}$ cm

11 $x^2 = 2.5^2 + 7^2$
    $x^2 = 6.25 + 49$
    $x^2 = 55.25$
    $x = \sqrt{55.25} = 7.43\ldots = 7.4$ m (1 d.p.)

Section 4: Algebra

1 . Symbols and sequences

1 a $3x^2 + 6x^3$
    $= 3x^2(1 + 2x)$
  b $4x(x^2 + 5) - 3(2x - 4)$
    $= 4x^3 + 20x - 6x + 12$
    $= 4x^3 + 14x + 12$

2 a $3x - 2y - x + 3y$
    $= 2x + y$
  b $4x^3y^2 \times 2xy^5$
    $= 8x^4y^7$
  c $\dfrac{12x^2}{3x^6}$
    $= \dfrac{4}{x^4}$

3 a Triangle sequence: 6, 10, 14, …
    First term $= 6$, difference $= 4$
    nth term $=$ difference $\times n +$ (first term – difference)
    $= 4 \times n + (6 - 4)$
    $= 4n + 2$
  b $n = 20$
    Number of triangles $= 4 \times 20 + 2 = 82$

2 Equations

1 a $v = -6 + (2.4 \times 1.9)$
    $v = -6 + 4.56$
    $v = -1.44$
  b i $5x - 3 = 2 + 3x$
    $2x - 3 = 2$
    $2x = 5$
    $x = 2.5$
  ii $\dfrac{12 - x}{2} = 6.5$
    $12 - x = 13$
    $12 = 13 + x$
    $-1 = x$
    $x = -1$

  iii $3(2x + 1) = 7 - 5x$
    $6x + 3 = 7 - 5x$
    $11x + 3 = 7$
    $11x = 4$
    $x = \frac{4}{11}$

2 a $(3 \times 2) + (2 \times -5)$
    $= 6 + -10$
    $= -4$
  b $\dfrac{2 - (2 \times -5)}{3}$
    $\dfrac{2 - (-10)}{3}$
    $= \dfrac{12}{3}$
    $= 4$

3 $x + b = an$
    $\dfrac{x + b}{n} = a$
    $a = \dfrac{x + b}{n}$

4

x	$x^2 - x$	Comment
3	$9 - 3 = 6$	Too low
4	$16 - 4 = 12$	Too high
3.5	$12.25 - 3.5 = 8.75$	Too low
3.6	$12.96 - 3.6 = 9.36$	Too high
3.55	$12.6025 - 3.55 = 9.0525$	Too high

$x = 3.5$ (1 d.p.)

5 $5x - 11 < 2x + 1$
    $3x - 11 < 1$
    $3x < 12$
    $x < 4$

3 Coordinates and graphs

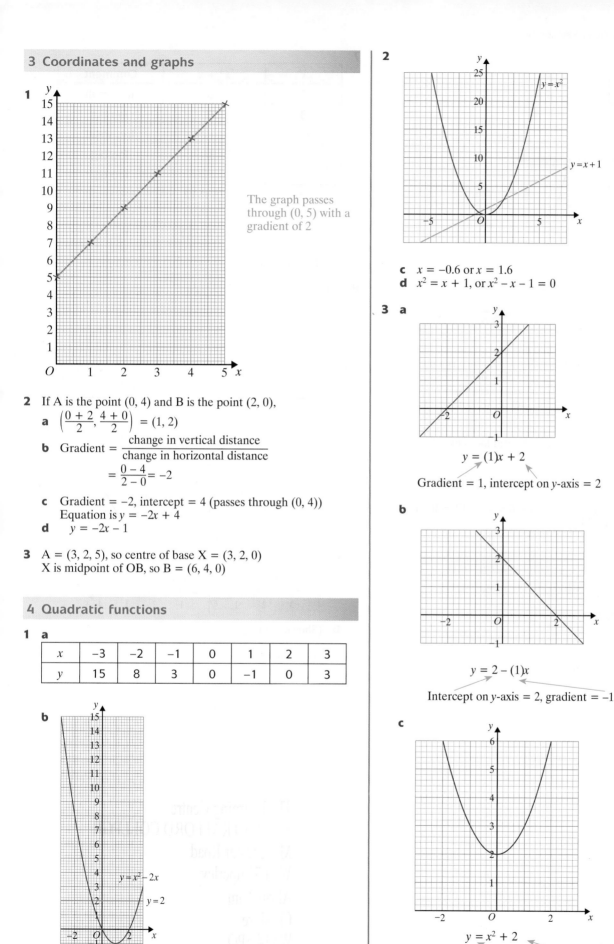

1

The graph passes through (0, 5) with a gradient of 2

2 If A is the point (0, 4) and B is the point (2, 0),

a $\left(\dfrac{0+2}{2}, \dfrac{4+0}{2}\right) = (1, 2)$

b Gradient = $\dfrac{\text{change in vertical distance}}{\text{change in horizontal distance}}$

$= \dfrac{0-4}{2-0} = -2$

c Gradient = -2, intercept = 4 (passes through (0, 4))
Equation is $y = -2x + 4$

d $y = -2x - 1$

3 A = (3, 2, 5), so centre of base X = (3, 2, 0)
X is midpoint of OB, so B = (6, 4, 0)

4 Quadratic functions

1 a

x	−3	−2	−1	0	1	2	3
y	15	8	3	0	−1	0	3

b

$y = x^2 - 2x$

$y = 2$

c $x = -0.7$ and $x = 2.7$

2

$y = x^2$

$y = x + 1$

c $x = -0.6$ or $x = 1.6$

d $x^2 = x + 1$, or $x^2 - x - 1 = 0$

3 a

$y = (1)x + 2$
Gradient = 1, intercept on y-axis = 2

b

$y = 2 - (1)x$
Intercept on y-axis = 2, gradient = -1

c

$y = x^2 + 2$
∪-shaped (x^2), intercept on y-axis = 2

d

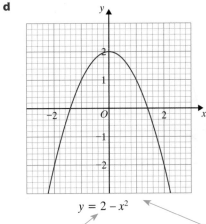

$y = 2 - x^2$

Intercept on y-axis = 2, $\cap$ -shaped ($-x^2$)

5 Proof

1 Let the numbers be $n, n + 1$ and $n + 2$
$n + n + 1 + n + 2 = 3n + 3 = 3(n + 1)$, which is a multiple of 3.

2 No, 2 is a prime number, but the sum of two consecutive numbers is always odd.

3 Call the top left-hand number n; the grid looks like this:

n	$n +1$
$n + 5$	$n + 6$

The sum is $n + n + 1 + n + 5 + n + 6 = 4n + 12 = 4(n + 3)$, which is a multiple of 4.

Section 4: Algebra

ANSWERS TO EXAMINATION STYLE QUESTIONS

1 $(2g^2h^5)(3g^3h) = 6g^5h^6$

2 $6ab - 4b = 2b(3a - 2)$

3 Difference is 4; nth term is $4n - 1$

4 $2(3a - 1) = a + 8$
$6a - 2 = a + 8$
$5a = 10$
$a = 2$

5 $p = 4r + t$
$p - t = 4r$
$\dfrac{p - t}{4} = r$

6 $s = ut + \frac{1}{2}at^2$
$s = -2 \times 4.2 + \frac{1}{2} \times 9 \times 4.2^2$
$s = -8.4 + 79.38$
$s = 70.98$

7

x	$x^3 + x$	Comment
2	10	Too small
3	30	Too big
2.5	18.125	Too small
2.7	22.383	Too big
2.6	20.176	Too big
2.58	19.753512	Too small
2.59	19.963979	Too small
2.595	20.069794875	Too big

$x = 2.59$ (2 d.p.)

8 C is $\left(\dfrac{4 + 2}{2}, \dfrac{-1 + 4}{2}\right) = (3, 1.5)$

9 B is (2, 0, 5)
D is (6, 5, 0)
F is (8, 2.5, 2.5)

10 a

x	-4	-3	-2	-1	0	1	2	3	4
x^2	16	9	4	1	0	1	4	9	16
$y = x^2 + x$	12	6	2	0	0	2	6	12	20

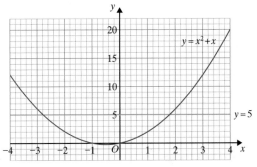

b (Shown in blue)

c (1.8, 5) and (−2.8, 5)

11 a Q(P + 1) is even × (odd or even), which is always even.

b P − Q is (odd or even) − even, which could be odd or even.

The Learning Centre
SOUTH TRAFFORD COLLEGE
Manchester Road
West Timperley
Altrincham
Cheshire
WA14 5PQ